Prawns of Japan and the World

(Translated from Japanese)

Chiaki Koizumi

(Editor-in-Chief)

ISBN 90 5410 769 3

A.A. Balkema, P.O. Box 1675, 3000 BR Rotterdam, Netherlands
Fax: +31.10.4135947; E-mail: balkema@balkema.nl
Internet site: http://www.balkema.nl

Distributed in USA and Canada by
A.A. Balkema Publishers, 2252 Ridge Road, Brookfield, Vermont 05036, USA
Fax: 802.276.3837; E-mail; Info@ashgate.com

Foreword

The marine resources along the coast and offshore Japan cover an area of about 200 nautical miles. It is assumed that the marine catch available to the Japanese people is about 6 million tons. This figure was calculated on the basis of a survey carried out by the Resource Survey Committee of the Prime Minister's Office, Government of Japan. At present (1982), this amount includes about 3 million tons of pilchard.

Marine products provide half of the animal proteins (about 25% of the total protein is supplemented) required by the Japanese people. The 6 million tons of marine products, including sardines, constitute a gift from Nature as a food resource. However, the feeding habits of the modern-day Japanese population do not always comply with the calculated domestic supply. The people of Japan, compared to most peoples in the world, in particular those of developing countries, lead an affluent life that increasingly desires high-quality products, including fishes and other marine products. To meet this demand, Japan imports large quantities of marine products from other countries.

Prawn is a major item among the imported marine products and a large sum of money is spent on its importation to satisfy the varying appetites of people in terms of texture and flavor. Though the world "prawn" is popularly used, it in fact covers a wide range of shrimp and allied crustaceans, which vary greatly in taste. Although several species of prawns are caught in Japanese waters, to supplement the marine catch, i.e., natural resources, aquaculture is intensively practiced as well. Japanese methods and equipment for prawn fishing are widely used abroad.

Scientific data relating to all these aspects of prawns were collected and included in an Open Lecture Program held at Tokyo University of Fisheries. Many experts participated and their wonderful cooperation is deeply appreciated.

April, 1984   Yoshino Amano, Director
                                            Tokyo University of Fisheries

Preface to this Edition

This book was originally published as a compilation of the lectures on "Prawns of Japan and the World" presented at the Ninth Extension Lecture Program of the Tokyo University of Fisheries, held in the summer of 1982.

Published in June, 1984, slightly more than a year after completion of the Extension Lectures, many queries were received from enthusiastic readers regarding prawn feed, in particular of cultured prawns. While planning the Ninth Extension Lecture Series, we had indeed thought of including a lecture on the feed of cultured prawns. But since these lectures were intended for a wide circle of participants, including consumers, the course had to be so designed as to minimalize difficult-to-understand technical jargon. So, the lecture on feed for prawns was excluded.

Now, in revising this book, a separate chapter on "Feed for Prawn Culture" has been added, which is actually an amalgamation of lectures designed for the extension program. This chapter was written by Prof. A. Kanazawa of the Faculty of Fisheries, Kagoshima University, and we are very grateful for his contribution.

We eagerly await the queries and comments of our readers in respect of this edition.

Chiaki Koizumi
Chairman, Ninth Extension Lecture Program
Tokyo University of Fisheries

Preface to the 1st Edition

The total annual supply of prawns in Japan, which includes headless prawns and peeled prawns, is said to have reached 200,000-220,000 tons in the early 1980s. Only a small percent is exported since most of the production is consumed within the country. The ratio of prawn catch in the total marine catch of the world (1,500,000-1,600,000 tons) is very large. Details of the catch indicate that production within the country is only about 50,000-60,000 tons; the remainder is imported from other countries. The amount of imported prawns is not at all small. In terms of money, the cost is about 300 billion yen. This is extremely large in terms of the ratio covered in the total amount of imported marine products. Japan is the foremost prawn-consuming country in the world and concomitantly holds first place in prawn importation.

The Japanese people, although they consume such a large quantity of prawns, often lack proper knowledge about them. There are many species distributed in the world seas and the species consumed as food are likewise many. However, when large quantities of prawns are imported from various parts of the world, they are handled under a few trade names such as pink prawns, white prawns, tiger prawns, brown prawns, etc. One wonders whether the common consumer has sufficient knowledge to enable him to differentiate *kurumaebi* and *taishoebi* when they are displayed in the market.

The total catch of prawn in the world is said to have reached maximum and efforts are underway to develop new fishing grounds for deep-sea prawns. The current position and the future prospects of the trends of prawn resources and aquaculture are major concerns of the people involved in the culturing and distributing (marketing) of prawns.

In view of the present status quo, the ninth extension lecture program on prawns of Japan and the world was convened at Tokyo University of Fisheries in 1982. The objective of this program was to provide the general consumer and those involved in the marketing, aquaculture, and processing of prawns with a better knowledge of these crustaceans. This book is a compilation of the lectures delivered during this program; the experts were requested to submit an edited manuscript of their presentations. During the course of planning the extension lecture program, it was disturbing to realize that few of the experts per se had actually carried out research on prawns. Therefore the participants in the program could not provide comprehensive details of the transition and future trends relating to worldwide prawn resources. Despite this drawback, the discussion on prawns covers a wide range of topics, such as species of prawns identified throughout the world, their ecology, the current position and future prospects of aquaculture in Japan and other countries, fishing implements, fishing methods, various problems relating to chemical aspects, use, processing, distribution and human consumption of prawns, and the history of coastal fishing for prawns in Japan. I express my sincere gratitude to all the participants in the program for their generous cooperation in providing valuable information.

The lecture program included participants ranging from the common consumer to professionals. Certain parts of the book may be slightly difficult for the layman to understand. However, since books pertaining to the latest information on prawns are few and not readily available, it is hoped that this book will prove useful not only for general information about prawns, but as a reference and classroom textbook.

April, 1984   Chiaki Koizumi
                        Chairman, Ninth Extension Lecture Program
                                     Tokyo University of Fisheries

Contents

List of Contributors

A. Kanazawa
Nutrition Chemist
Marine Resource Center
Kagoshima University

K. Kanda
Prof. Fishery Metrology (Retired)
Tokyo University of Fisheries

F. Nagayama
Prof. Food Chemistry
Tokyo University of Fisheries

A. Nakai
Prof. Dept. Economics
Fuji University

Y. Ogasawara
Prof. Emeritus
Tokai University of Fisheries

N. Sako
Rerearch Scholar

M. Takeda
Head, Zoological Research Section No. 3
Tokyo University of Fisheries

Y. Uno
Prof. Emeritus
Tokyo University of Fisheries

S. Wada
Assoc. Prof. Food Preservation
Tokyo University of Fisheries

T. Wakamatsu
Aquaculture Consultant
Tokyo University of Fisheries

Shrimps and Prawns of the World

Shrimps and prawns are well known as important marine products. A major social issue was raised by herring roe as the "yellow diamond" and now a similar issue appears during the course of negotiations for prawns and shrimps. Prawns and shrimps are imported by Japan in large quantities from various countries of the world. When displayed in the market, however, all are referred to as either *kurumaebi* or *taishoebi*. There have been problems in the sale of many unfamiliar species of fishes, e.g. yellowtail and species of bivalves, e.g. little clams or hard clams. The large species of prawns and shrimps need detailed systematic scrutiny for proper identification.

The word *"ebi"* calls to mind the *"kurumaebi"* or prawns (*Penaeus japonicus*) and the expensive spiny lobsters called *"ise-ebi"* in Japanese. These names refer to the crawling and swimming types respectively and in Japan, the former is grouped under *ebi* and the latter under *kairo*. This type of classification indicates the long association between prawns, Shrimps, and man. In English (American and British) also terms such as prawn, shrimp, and lobster are used. Prawn and shrimp refer to crawling types while lobster refers to the swimming type. In Britain those with a body length of more than 5 cm are called prawns and the smaller ones termed shrimps. In America there is no such differentiation and they are mostly clubbed together as shrimp. Such discrepancies in American and British terminology reflect the formal nomenclature of FAO and indicate a conflicting trend. For example, while all species of *Palaemon* are called prawns and most of the others shrimps, *Penaeus* of the eastern and western coasts of America are called shrimps and the others prawns.

Lobsters are distinguished into European and American lobsters and rated high as seafood. Spiny lobsters are called *ise-ebi* in Japanese. Given the progress in transportation techniques, a rich variety of prawns and shrimps from other countries is available in Japan. In fact, varieties of spiny lobsters of the Southern Hemisphere are commonly found in the market. These are casually referred to as crayfish. Since the term crayfish is generally applied to many species of lobsters in Europe and America, the species imported by Japan is known as *minami-ise-ebi* or southern lobster, and was eventually named rock lobster in 1969.

Crayfish belong to the crawling type (Reptantia) and since they differ distinctly in appearance from other prawns and shrimps, in Japan they are called *riko*. In Europe they are called crayfish and in America, crawfish.

Besides the scientific name, each species has a standard label name from its country of origin and many have regional names as well. For those circulated in the trade industry, there are trade names based on body color or place of production. In this book, standard English names or FAO recognized names are given in addition to scientific names.

The popular literature on prawns and shrimps includes the book titled *Ebi*, a Suisansha publication written by Noboru Sakamasu. This text mainly deals with the commercial types of prawns and shrimps and those which have potential as a future source. Owing to constraints of space, it is not possible to cover all the types in detail but the literature and illustrations give an idea of the types otherwise not covered in descriptions.

1.1   MORPHOLOGY AND CLASSIFICATION OF PRAWNS AND SHRIMPS

As explained above, prawns and shrimps can be classified into the swimming type (Natantia) and crawling or creeping type (Reptantia). Though there is some difference in degree of development, all prawns and shrimps have a hard exoskeleton (test, carapace, commonly "shell"), a well-developed abdomen, and long antennae. Given these common features, a comparison of prawns and shrimps with common crabs and hermit crabs, closely related to them, immediately sets the former apart from the latter. The same features likewise readily distinguish prawns and shrimps from krills and mysids also.

1.1.1   Morphological Characteristics

The body of prawns and shrimps is bilaterally symmetrical and segmented. The head is composed of 5 segments, the thorax of 8 segments, and the abdomen of 7 segments. The head and thorax are covered by a common exoskeleton called the carapace. Depending on the species of prawn, the carapace has supraorbital spines, pleuroorbital (lateral) spines, supraantennal spines, branchiostegal spines, and hepatic spines. In addition to spines, folds and grooves occur in the carapace of many species. At the anterior end the carapace is extended into a rostrum. The length, angle, presence or absence of serrations, and their number serve as important features for a detailed classification.

Each segment of the cephalothorax as well as the abdomen has a pair of appendages, modified according to their location, having different functions. The 5 pairs of appendages of the head from the anterior end are the 1st antennae, 2nd antennae, mandibles, maxillae I, and maxillae II. The 1st antennae are short and serve as chemoreceptors. The 2nd antennae are long and serve as tactile receptors. The other 3 pairs of appendages are included in the mouthparts. The mandibles are in the form of strong teeth. Of the 8 pairs of thoracic appendages, the first 3 pairs are the maxillipeds I to III and are included in the mouthparts along with the mandibles, maxillae I and maxillae II of the head region. As for the other 5 pairs of appendages, some species, such as spiny lobsters, totally lack chelae (pincers) while others, such as prawns of the genus *Penaeus* have chelae in the first 3 pairs, some, e.g. prawns of the genus *Palaemon*, have chelae in the first 2 pairs, and some, e.g. lobsters of the genus *Cambarus*, have chelae in all 5 pairs. The formation of chelae and the variations in their morphology, as well as the presence or absence of the exopodite on the appendages form the basis for classification into genera or even families. The abdominal appendages are composed on an endopodite and exopodite (leaflike) and serve as locomotory organs. Some forms, such as prawns of the genus *Penaeus*, release their eggs in the water while those of the genus *Sergestes* retain and protect the eggs by keeping them attached to the abdominal appendages until the time of hatching. In the male, the first pair of abdominal appendages is modified into the copulatory organs. They also form the basis for classification. The terminal segment of the abdomen is called the telson and along with the appendages (uropods) of abdominal segment VI, it (the telson) forms the tail fan (Fig. 1.1).

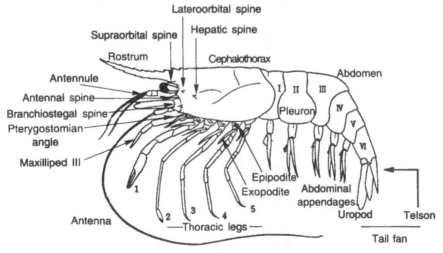

Fig. 1.1. External features of a prawn.

The male genital apertures open at the coxopodite (coxa) of the fifth thoracic appendages. In the female, the genital apertures are located in the coxopodites of the second thoracic appendages. Thus the male and female can be differentiated on the basis of location of the genital aperture. For this purpose, the feature relating to the size of the chelae is not very reliable.

1.1.2 Mysids and Krills

Besides the true shrimps and prawns there are many crustaceans which are casually referred to as prawns or shrimps. The so-called prawn phase with well-developed abdomen is close to the original form of the crustacean and among the lower crustaceans all the basic forms which have not undergone specialization belong to the prawn phase. The name prawn, *ebi* in Japanese, is associated with several groups, namely, *honenebi* (Branchinella), *kabutoebi* (Apus), *kaiebi* (Caenesthriella), *konohaebi* (Nebalia), and *yokoebi* (Gammarus). Each group has innumerable varietal names. However, all the species are small and lack the typical prawnlike carapace and therefore are readily differentiated. On the other hand, there are groups called mysids and krills which have also been referred to as *ebi* sometimes. Unless one is familiar with mysids and krills, it is difficult to distinguish them from prawns and shrimps. Since many are used in the fishery industry, a good knowledge of their specific characteristics is desirable.

About 760 species of mysids have been identified throughout the world and new species are constantly being recorded. Mysids differ from prawns in the following features: posterior segments of the thorax protrude from the carapace; the 8 pairs of thoracic appendages are almost identical and all are composed of an exopodite and an endopodite; the endopodite of the uropod has a statocyst (balancing receptor) and the female develops a brood pouch in the thoracic part. Several eggs are held in the brood pouch where they undergo full development and the hatchlings emerge swimming. In natural conditions they are important feed for fishes. Freshwater *Neomysis intermedia*, spring-water *N. japonica*, and marine *Anisomysis ijimai* are caught in large numbers and used in the preparation of a Japanese dish known as *tsukudani* (precooked food boiled in soybean soup), or as fish feed, or as fishing bait.

Eleven genera and 85 species of krills have been reported throughout the world. Though the number of species is small, the overall quantity of the organisms is very large. They serve as feed for baleen whales and waterfowl and play an important role in the circulation of organic substances in seas and oceans. The krill identified as *Euphausia superba* has recently drawn much attention as a new organic resource for human beings. It is a large species among krills. The body length is more than 5 cm and the resource is said to be about several hundred million tons. Hence it cannot be ignored in future. Further, fishing for *E. pacifica* in spring along Miyagi coast is also expected to yield good results thanks to the angling boom in Japan.

Krills strongly resemble prawns. However, all the 8 pairs of thoracic appendages are composed of an endopodite and an exopodite and cannot be differentiated into mouthparts and chelated legs. One of the chief characteristics is the presence of branchiae at the base of the thoracic legs.

Besides mysids and krills there exists the genus *Acetes*. Its members are used for pickling, fishing bait, and feed in pisciculture. *Acetes* are a type of shrimp belonging to the family Sergestidae.

1.1.3 Systematic Position

Prawns and shrimps are classified as phylum Arthropoda, class Crustacea, subclass Malacostraca, superorder Eucarida, order Decapoda and suborder Macrura. Prawns and shrimps are popularly called macrurans after the name of the suborder, which means long-tailed. Thus Macrura stand in contrast to deformed tailed forms such as hermit crabs and short-tailed forms such as crabs. Another classification includes the swimming forms under the suborder Natantia and the crawling prawns, hermit crabs, and crabs under the suborder Reptantia. These classifications are simple and easy but intensive studies are currently being carried out on details of the gills, protection and preservation techniques of the eggs, and variations in the period of hatching. According to this system the order Decapoda is divided into 2 suborders, namely, Dendrobranchiata and Pleocyemata. The members of the former order have dendrobranchiae composed of branchial filaments. From the eggs released in the sea, nauplius larvae hatch. The families Sergestidae and Penaeidae belong to this order. Today the subfamilies earlier included under the family Penaeidae have been elevated to the status of family, thus making a total of 3 families, namely, Penaeidae, Aristeidae, and Sicyonidae (see Table 1.1).

Except for the 350 species belonging to the suborder Dendrobranchiata, all the 2,500 species of decapods belong to the suborder Pleocyemata. In these, the branchial filaments are not branched and the larvae hatch at a more advanced stage than nauplius and are held attached to the abdominal appendages of the female parent. According to this system, hermit crabs and crabs should be included under the suborder Dendrobranchiata.

Nowadays there is a strong move to divide the order Decapoda into 4 suborders. These include the suborder Dendrobranchiata and the suborder Reptantia referred to earlier, and Euzygida with just one family (Stenopidae) and Eukyphida with the other. This division appears to be based on the earlier concept of Natantia (swimming forms) and Reptantia (crawling forms).

Table 1.1: Classification of prawns and shrimps

Phylum Arthropoda————————————————————Order Decapoda
Class Crustacea   Subclass Malacostraca
Superclass Crustacea   Class Malacostraca   Subclass Eucarida

Macrura			
Natantia			Reptantia
Dendrobranchiata	Euzygida	Eukyphida	
	Pleocyemata		

Penaeida	Stenopids	*Koebi* (Caridae)	Lobsters
*Fam. Sergestidae	F. Stenopidae	F. *Teppoebi*	F. Cambaro
*Fam. Aristeidae		F. *Tsunome-ebi*	F. Cambaridae
*Fam. Solenoceridae		F. *Moebi*	F. *Minamizarigani*
*Fam. Sicyonidae		F. *Rosokuebi*	F. *Okazarigani*
*Fam. Penaeidae		F. *Mikawaebi*	F. Nephropidae
		F. *Sarasaebi*	F. *Osate-ebi*
		F. Bresiliidae	
		Crawfish	Rock lobsters
Note: Japanese species are not		F. Oplophoridae	F. *Senjuebi*
available under taxa given		F. Nematocarcinidae	F. Glypheidae
in English		F. Atyidae	F. Palinuridae
		F. Pasiphaeidae	F. Scyllaridae
*Taxa marked with an asterisk		F. Palaemonidae	F. Synaxidae
include species important for		F. Gnathophyllidae	
commercial purposes. They are		F. Campylonotidae	
detailed in this book		F. Pandolidae	
		F. *Okinagare-ebi*	
		F. *Ukikabutoebi*	
		F. Procarididae	
		F. *Igoguriebi*	
		F. Stylodactylidae	
		F. Crangonidae	
		F. Glyphocrangonidae	

Notwithstanding the classification of the suborder and the nomenclature, it is clear that decapods comprise swimming and crawling forms. The penaeids vary widely in characters and *Stenopus (Otohime-ebi)* exist along with other small genera.

1.2   SPECIES IMPORTANT FOR FISHERIES

All kinds of prawns and shrimps are potential food sources. However, the question is: to what extent are they valuable as commercial products. This depends on the quantum of the resource and the size of the individual form. The natural resources of shallow seas should be developed further, but without depletion! and bearing this in mind efforts should be directed to a fish culture program. Exploitation of new resources (deep sea) is also hoped for in the near future. Prawns of the genus *Penaeus* *(Kurumaebi)* have been successfully cultured and many more species are expected to follow suit shortly. Surveys of the deep sea as a new resource have proven extremely expensive, however, both in terms of fuel and manpower. The economic viability of this resource must therefore be given careful consideration.

The prawns and shrimps targeted for commercial catch, those which constitute the bulk of the catch, and those which appear to be potential new resources are broadly discussed below. Analysis of the catches showed that the species which can be used fresh/raw are few. Therefore development of the processed-fish industry is strongly advocated.

1.2.1 Penaeids (*Kuruma* Shrimp)

Almost all types of penaeids are important for fisheries. Therefore intensive studies have been carried out on their taxonomy, ecology, development, and other aspects and impressive data compiled. Recent studies on American forms have drawn much attention and according to modern systematics, the priorly recognized family Penaeidae has been split into four families, namely, Penaeidae, Aristeidae, Solenoceridae, and Sicyonidae. These are classified as follows.

1. **Family Sergestidae**....4th and 5th thoracic legs extremely short. These are normal in family Aristeidae.
2. **Family Aristeidae**....Inner flagellum of 1st antenna very short. More or less same length as outer flagellum in family Solenoceridae.
3. **Family Solenoceridae**....Flagellum of 1st antenna cylindrical. Whiskery in family Sicyonidae.
4. **Family Sicyonidae**.....Abdominal appendages lack endopodite. Latter present in family Penaeidae.

(1) Family Sergestidae

Sakuraebi or *sakura* shrimp, identified as *Sergia lucens*, is a very popular food item throughout Japan. It is also known for its iridescence and solar cycle of vertical migration. In spite of its limited distribution in Tokyo Bay, Sagami Bay, and Ayakawa Bay, amazingly the annual catch exceeds 5,000 tons. About 35 species of *sakura* shrimps have been recorded in various parts of the world. However, only the Japanese species is the object of commercial fishing. No other planktonic organism is used as a resource to the extent that *Sergia lucens* is.

On the other hand, *Akiami* paste shrimp *Acetes japonicus* is distributed up to the western part of the Indian Ocean. It is caught by several countries. From the Pacific Ocean of the West Indies, 10 species and 2 subspecies, from the East Pacific Ocean 1 species, and from the West Atlantic Ocean 3 species and 1 subspecies have been identified. It is a small shrimp with a maximum body length of 3 cm but is an important food item in various parts of Southeast Asia. In particular, the *Jawla* paste shrimp *A. indicus*, distributed between Indonesia and India, is caught in large numbers and is said to constitute 20% of the total shrimp catch in Bombay.

(2) Family Aristeidae

The Japanese equivalent for Aristeidae, that is *Chihiroebi*, transcribed into Chinese characters, means shrimps of the deep sea. As deep-sea free-swimming shrimps, genus *Gennadas* is very popular. However, measuring about 5 cm in body length, they do not merit commercial value as a marine product.

The larger species, more than 20 cm in length, are found in 3 genera, viz., *Aristeomorpha*, *Aristeus*, and *Plesiopenaeus* and a few species have commercial value. On the whole, though, they are not currently primary targets due to the fact that the palatable content is small, the fat content high, and tremendous effort is required to catch them as they live in deep seas.

Genus *Aristeomorpha*: The giant red shrimp* *A. foliacea*, caught at depths of 200-400 m in Sagami Bay, Enshu Strait, Kumano Strait, and the southwestern sea region of Kyushu, is widely distributed in the Indian and Atlantic oceans. It is caught in several countries other than Japan. The production rate is not large, however. The depth of the habitat has been recorded as 250-1,300 m. It appears to be almost a benthic dweller. The Indian red shrimp *A. woodmasoni* is also caught at depths of 330-500 m but on a small scale.

Genus *Aristeus*: The stout red shrimp *A. virilis*, which bears a large number of luminescent organs in the thoracic legs, measures 19 cm in body length. It dwells at depths of 350-800 m and is distributed up to the western region of the Indian Ocean. It is considered a hidden resource as is the Arabian red shrimp *A. alcocki*. In the Mediterranean Sea, the blue and red shrimp *A. antennatus* and in the sea off the western coast of Africa, the striped red shrimp *A. varidens* are caught in large numbers. Both species live at depths of 200-1,200 m and are caught by trawlers at depths of 400-600 m.

Genus *Plesiopenaeus*: The scarlet shrimp *P. edwardsianus* (Fig. 1.2), measuring more than 30 cm in body length, is distributed at depths of 275-1,850 m, mainly 400-900 m in the East and West Atlantic Ocean. Spanish ships operate from Senegal to Guinea and from the Congo to Angola. The catch is frozen in the ships and sold in Spain and France. According to the survey carried out by the Marine Fisheries Resource Development Center in the offseas of Surinam and French Guiana in the West Atlantic Ocean, large numbers were caught and an experimental

---

*Commonly known as the blood-red shrimp—Technical Editor.

Fig. 1.2. Scarlet shrimp.

sale done inside the countries, which was only moderately successful. The genus includes 2 species. The other species, *P. armatus*, is also found in Japanese waters. It is a large deep-sea dweller widely distributed even in the Atlantic Ocean but an estimation of its resources has yet to be done.

(3) Family Solenoceridae

According to latest studies, this family contains 7 genera, of which the 4 mentioned below include species used as marine products.

Genus *Haliporoides*: Two species are recorded from the West Pacific Ocean and 1 species from the East. All three are of high commercial value as marine products and measure about 20 cm in body length. The species *H. sibogae*, popularly known as jackknife shrimp, is highly abundant even in Japanese waters and densely populous from the Malay archipelago to Australia and New Zealand. Its importance is recognized even in Madagascar. Its habitats range from 100-1,460 m depths and catching operations are carried out at depths of 350-600 m. The rostrum is short but the antennae are long; hence it is called the long-whiskered shrimp (*Higenagaebi*) in Japanese. The knife shrimp *H. triarthrus*, which has already acquired an important position in South American waters, lives on the bottom at depths of 350-450 m. Since it is rather small, about 15 cm in body length, it is mostly sold as peeled shrimp.

The Chilean knife shrimp *H. diomedeae*, found at depths of 300-1,800 m from Panama to Chile, has yet to be explored. However, considering the regions with high habitat density, this potential resource may soon be recognized.

Genus *Pleoticus*: Two species from the West Atlantic Ocean and one species from the Red Sea have been recorded. The two species which have drawn the attention of fisheries are the Argentine red shrimp *P. muelleri* distributed from southern Brazil to Argentina and the royal red shrimp *P. robustus* distributed from Massachusetts (USA) to French Guiana. The Argentine shrimp dwells on a sandy bottom at depths of 5-25 m, water temperature 10-23°C, and chlorinity 33.27-33.94%. The average body length is 10 cm and the maximum 19 cm. It constitutes the major part of the shrimp catch of Argentina. The annual catch is 200 tons (minimum) and most of the catch taken between 41° S and 44° S. Efforts are underway to culture these species in Argentina.

As for *P. robustus*, its habitat at depths of 250-750 m posed a problem. In the early 1950s, experiments were carried out and after 1962 a commercial catch initiated in the waters around peninsular Florida. In 1975, the catch in America was 122 tons, increasing to 136 tons in 1976. The catch is considerably rich in the northeastern waters of South America.

Genus *Solenocera* possesses the typical features of the family Solenoceridae. The inner and outer flagella of the 1st antennae are more or less equal in length and about 1.5 times the carapace length. The habitat depth varies but it usually dwells on a muddy bottom. Hence it is called the mud shrimp. The shallow-water types are catch targets in several countries but it is not as important as *Kuruma* shrimp. This is because the catch is small and the abdomen strongly flexed with little flesh. There are 6 species in Japan and the ridgeback shrimp (*S. choprai*) is sometimes packed and transported. However, the catch does not specifically target this species.

Except for the coastal mud shrimp *S. crassicornis* distributed in shallow waters between the Malay archipelago and Pakistan, the *Kolibri* shrimp *S. agassizii* of the East Pacific Ocean and the Atlantic mud shrimp *S. membranacea* of the East Atlantic Ocean are caught on a commercial basis; other species, like the Japanese species, are not. Regarding *S. vioscai* of the Gulf of Mexico and *S. acuminata* distributed in the southern part of the Caribbean Sea, experimental catches carried out after 1950 indicated that the catches are generally poor and unstable. Their commercial value is low and hence catches cannot be raised to a commercial level, although they might be if the catch rate could be stabilized.

(4) Family Sicyonidae

The family includes only one genus, *Sicyonia*. In Japanese it is called *Ishiebi*, meaning rock shrimp, which is also the English name. True to its name the carapace is stone hard. Since most of the species are not even 10 cm in body length, they merit little attention from fisheries. None of the 8 Japanese species is commercially viable. The West Atlantic rock shrimp *S. brevirostris* is exceptionally large at 15 cm in length as well as tasty. It is caught in the waters between North Carolina and Florida. The annual catch in 1975 reached 900 tons and fishing was later extended to Mexican coasts.

The lesser rock shrimp *S. dorsalis* and the Kinglet rock shrimp *S. typica* distributed from North Carolina (USA) through the Caribbean Sea up to Brazil are small. They are abundant in shallow waters and can withstand commercial fishing. Species of the East Atlantic Ocean and West Pacific Ocean are used regionally.

(5) Family Penaeidae

The very mention of the world *"ebi"* calls to mind the species of *Penaeus* without which a shrimp catch would not materialize. The family Penaeidae contains 14 genera and 120 species. Of these, the following are objects of commercial fishing: genus *Artemesia* 1 species, *Atypopenaeus* 2, *Acropetasma* 1, *Metapenaeopsis* 17, *Metapenaeus* 23, *Parapenaeopsis* 15, *Parapenaeus* 7, *Penaeopsis* 2, *Trachypenaeus* 12, *Xiphopenaeus* 2, and *Penaeus* 28, totaling 110 species. Of the 1,700 species of small shrimps, constituting the majority of prawns and shrimps, only about 30 are used commercially. Hence it should be conceded that mainly members of the genus *Penaeus* support the prawn and shrimp catch.

Genus *Artemesia*: Includes only one species, *A. longinaris*, the Argentine stiletto shrimp. Its range lies between Brazil and Argentina. The body reaches a maximum length of 15 cm. It dwells in a sandy mixed bottom at a depth of about 60 m. It is caught together with *Pleoticus muelleri* of the family Solenoceridae mentioned earlier. The annual catch in Argentina is 250-500 tons and efforts are underway to culture it.

Genus *Atypopenaeus*: The periscope shrimp (*A. stenodactylus*) caught in Setionaikai of Japan is a small species about 5 cm in body length. It is caught around the coastal West Indies. The orange shrimp *A. formosus* ranging between New Guinea and the eastern coast of Australia is also caught but is of low commercial value.

Genus *Macropetasma*: An exclusive genus of south and southwest Africa that includes only one species, *M. africana*, a swimming shrimp. It is a small species about 5 cm in body length and dwells at a depth of about 25 m near a river mouth. At present it is not fished even though prolific but is said to be a good natural resource.

Genus *Metapenaeopsis*: Of the 11 species in Japanese waters, the whiskered velvet shrimp *M. barbata* and Tora velvet shrimp *M. acclivis* are caught in large numbers in small dragnets along coastal regions. Other species are also caught and the dominant type varies in different regions.

As the English name indicates, the carapace is covered by short bristles and mandibular and hepatic grooves. While the antennal spines, preantennal spines, and hepatic spines are well developed, the supraorbital spines are small. In some species, vocal organs on short protuberances are present on the posterior part of the carapace. The whiskered velvet shrimp and Tora shrimp mostly live together. Live, the Tora shrimp looks red. The two species can be differentiated on the basis of number of protuberant vocal organs and size of the coxa of the 4th walking legs in females (in Tora shrimp, it is unusually large). However, external features alone do not suffice in differentiation.

On the whole, the species of this genus, except for the whiskered velvet shrimp and Tora shrimp, are not very important.

Genus *Metapenaeus*: An important genus after *Penaeus* in which about 25 species are known. Unlike *Penaeus*, it has not been found in the East Pacific Ocean nor in the Atlantic. Morphologically, it is differentiated by the presence of teeth only on the upper margin of the rostrum and the absence of an exopodite on the 5th thoracic legs.

In Japan, there are 4 species, namely, the greasyback shrimp *M. ensis*, *shiba* shrimp *M. joyneri*, *moebi* shrimp *M. moebi* and middle shrimp *M. intermedius*. The latter is caught at depths of 20-60 m in Tosa Bay but the number is small. The other 3 species are important in fisheries and many studies—biological and marine biological—have been carried out. The greasyback shrimp and *shiba* shrimp are found at a depth of 20 m and dwell on a mud bottom. Spawning takes place between mid-June and late September. Juvenile shrimps are found in spring-water basins also. By October they have grown to 2-3 cm in length and gradually emerge from the bay. They live for a year. In the case of prawns and shrimps (penaeids) of the coastal region, they begin to accumulate in groups while migrating to wintering locations. Catches target this period. Once the fish enter deep waters outside the bay, grouping disappears, except for *shiba* shrimp, which continue to remain in groups even in wintering places. Hence they are caught even in winter. The greasyback shrimp is not only caught in Japan, but also in various regions of Southeast Asia and in large numbers. In Malaysia and Thailand efforts are on to culture them. On the other hand, *shiba* shrimp is confined only to the extreme east; in Korea, the annual catch has increased to 2,000 tons.

Moebi shrimp dwell on a sand-mud bottom. Spawning occurs for a month during peak summer. The juvenile shrimp with a carapace 1 cm in length emerge in October. They are caught in Southeast Asia where culturing is also in progress.

Greasyback shrimp are widely used in various parts of the world. The important species include the Jinga shrimp *M. affinis* and yellow shrimp *M. brevicornis* distributed in the Arabian Sea from the Malay archipelago, Kadal shrimp *M. dobsoni* distributed from the Philippines to India, endeavour shrimp *M. endeavouri* of Australia, and the speckled shrimp *M. monoceros* of the Indian Ocean. The latter is caught in large numbers and used in various forms. It is an important species in all the regions of its distribution. It has been reported that it is likely to migrate through the Suez Canal, enter the Mediterranean Sea, and become a significant marine product.

Genus *Parapenaeopsis*: The 2 Japanese species, *P. tenella* (smooth shell shrimp) and *P. cornuta* (coral shrimp), can be spotted among the shrimp catch but are not considered important species. The members found in other countries also have the same status. Even the larger forms are restricted in size to 15 cm. Moreover, they are not caught in large numbers.

Genus *Parapenaeus*: Externally it possesses no features specific to it. However, careful observation of the carapace does show a pair of longitudinal lines, one on either side or on the dorsal side, and this helps to differentiate it from the genera *Metapenaeopsis* and *Metapenaeus*. The body length is usually 15 cm and the main habitat is at depths of 150–400 m. Often termed "hidden resource", the genus *Parapenaeus* is not widely used. At present, only the deepwater rose shrimp *P. longirostris* (Fig. 1.3) has been successfully exploited. This species is distributed from the Mediterranean Sea to Angola through Portugal and in the East and West Atlantic Ocean from Massachusetts (USA) to French Guiana. This shrimp grows to a length of 18 cm. It is an important species in the regions around the Mediterranean Sea coast and western coast of Africa. In the southern part of Guinea, catching operations are carried out at depths of 200-300 m and daily catches reach 1-3 tons.

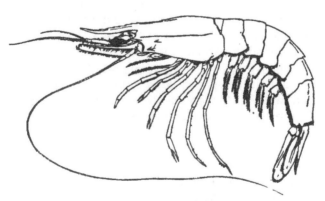

Fig. 1.3. Deepwater rose shrimp (*P. longirostris*) (FAO Identification Sheets, vol. 2).

Genus *Penaeopsis*: The needle shrimp *P. eduardoi* which dwells at depths of 180-750 m from Japan to the Indian Ocean has much potential as a biological resource. In Japan, it is caught in large numbers from the southern waters of Kyushu. The morphology and depth of habitat of the megalops shrimp *P. megalops*, distributed along both the eastern and western coasts of the Atlantic Ocean, closely resemble those of the needle shrimp. It too is a potential biological resource. During the survey carried out by the Marine Fisheries Resource Development Center a rich catch was made. Part of the catch was sold as "orange shrimp" because of its body color.

Genus *Trachypenaeus*: Represented by the southern rough shrimp *T. curvirostris*, which is small and of poor value. It enters small dragnets in large numbers in the coastal regions and supports shrimp fishing. Eighteen species are known throughout the world. Of these, 11 species are from the West Pacific Ocean, 5 from the East Pacific Ocean, and 2 from the Atlantic. All of them have a covering of short bristles. In Japan, the southern rough shrimp is caught in large numbers and there is greater demand for it as fishing bait. Surveys are underway in various regions. According to their reports, these shrimp live at depths of 20-30 m in the inner bay and at a depth of 50 m in places close to the open sea. The spawning season is May-October in western and southern Japan and July-September in Sendai Bay. Since the spawning period is long, the early spawners are large (8-10 cm) and the later spawners small (6.5-8 cm). Accordingly, there are long- and short-term generations. The longevity of large shrimps is said to be 2 years.

Genus *Xiphopenaeus*: Includes 2 species, namely, the Atlantic seabob *X. kroyeri* and the Pacific seabob *X. riveti* distributed in eastern and western central South America. From the point of view of fishing, the former species appears to be more important. Growth is restricted to 14 cm body length. The thoracic legs are extremely long and slender. The rostrum is long. The upper part of the orbit has 5 teeth. The species dwell at depths of 2-30 m and are found in particularly large numbers near river mouths (estuaries). The annual catch in the northern part of the Gulf of Mexico is 3,000 tons. They are also caught in large numbers in various regions of Brazil. It is even said that the seabob stabilizes the catching statistics of prawns and shrimps as a whole. However, due to its small size, its use is almost localized.

Seabobs on the Pacific Ocean side are distributed from Mexico to Peru but consumed only in certain regions.

Genus *Penaeus* (Fig. 1.4): Mainly lives in shallow waters of tropical and subtropical regions. To date, 28 species have been reported from various parts of the world and show similarity in morphological features. The genus differs from other genera of the family Penaeidae in having 1-4 teeth on the lower margin of the rostrum.

The carapace always has maxillary grooves, orbitoantennal grooves, antennal spines, and hepatic spines. The presence or absence of gastric sulcus and median sulcus depends on the species. The posterior half of

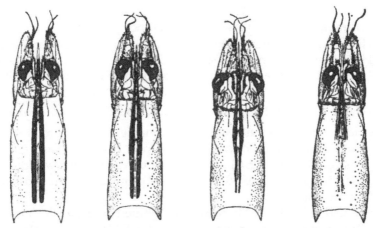

Fig. 1.4. Four types of *Penaeus* of central and southern regions of the West Atlantic Ocean. From left: pink-spotted, sazaan pink, sazaan brown, sazaan white. Note the rostral groove.

each abdominal segment has a median protuberance and the median protuberance of the 6th abdominal segment ends in a spine. Some species have a longitudinal groove on the middorsal surface of the caudal segment with movable spines on the two sides of the groove. The male genital organ is bilaterally symmetrical. The female reproductive organ in some species has a special structure known as the seminal receptacle while in other species there is no receptacle.

According to recent studies, on the basis of presence or absence of grooves or ridges on the carapace, its length, presence or absence of longitudinal grooves on the sides of the 6th abdominal segment, shape of gonads, geographic distribution and ecology, the genus is divided into 6 subgenera. These are *Penaeus* (3 species from the West Pacific Ocean), *Melicertus* (6 species from the West Pacific Ocean), *Fenneropenaeus* (5 species from the West Pacific Ocean), *Lithopenaeus* (5 species from the East Pacific Ocean and the Atlantic), *Marsupenaeus* (only *Kurumaebi* of the West Pacific Ocean), and *Farfantepenaeus* (8 species in the East Pacific Ocean and the Atlantic). The following features differentiate the subgenera.

1. Longitudinal groove present on lateral sides of abdominal segment VI .... 2.

— Longitudinal grooves absent on abdominal segment VI .... 3.

2. Groove or rostrum ends near hepatic spine. Gastric protuberance lacking. Female gonads of open type .... *Lithopenaeus*.

— Rostral groove extends to posterior margin of carapace. Gastric protuberance present. Female gonads of closed type ....
.... *Farfantepenaeus*.

3. Rostral groove short. Gastric protuberance lacking .... 4.
— Rostral groove long. Gastric protuberance present .... 5.
4. Hepatic protuberance lacking .... *Fenneropenaeus.*
— Hepatic protuberance present .... *Penaeus.*
5. Lateral plates of female gonads composed of right and left labium
.... *Melicertus.*
— Female gonads cylindrical .... *Marsupenaeus.*

P. (Fa.) aztecus: West Atlantic Ocean. FAO name, northern brown shrimp. More than 60,000 tons collected annually from North Carolina (USA) to the open seas off Texas (USA). It is the highest catch in the world. Found in high density at depths of 30-55 m. Males grow to 19.5 cm and females to 23.5 cm in length.

P. (Fa.) brasiliensis (Fig. 1.5): Widely distributed from North Carolina to Brazil and called the red-spotted shrimp by FAO. Locally often called pink-spotted or spotted-pink shrimp. Its unique feature is the presence of reddish-brown spots on the 3rd abdominal segment.

Fig. 1.5. Pink-spotted shrimp.

P. (Fa.) brevirostris: East Pacific Ocean from Mexico to northern Peru. FAO name, crystal shrimp. Body length 17 cm. Abundance moderate.

P. (Fa.) californiensis: California (USA) to northern Peru. Abundant at depths of 25-40 m. FAO name, yellowleg shrimp. Body length 20 cm. Caught in large numbers along coasts of various countries from Mexico to Central America. Catching rate in Ecuador has recently increased.

P. (Me.) canaliculatus: FAO name, witch prawn. Widely distributed from Taiwan to the Indian Ocean. Small, body length about 12 cm. Importance low. Because of its body color, often confused in the past with *Kurumaebi*.

P. (Fe.) chinensis: Standard local (Japanese) name *koraiebi* or *taishoebi* FAO name, fleshy prawn. Scientific name for many years, *P. orientalis*. In spite of being restricted to the Yellow Sea, its population is large.

P. (Fa.) duorarum: Important species distributed from the southeast coast of North America to the Gulf of Mexico. FAO name, northern pink shrimp. Large, body length 28 cm. Mostly lives at depths of 15-35 m. Important species in the Gulf of Mexico; annual catch exceeds 20,000 tons.

P. (P.) esculentus: FAO name, brown tiger prawn. Australian species distributed along west, north, and east coasts. Body length 15 cm.

P. (Fe.) indicus: FAO name, Indian white prawn; Japanese name *indoebi*. Body length exceeds 20 cm. Widely distributed from the Philippines to the western part of the Indian Ocean. Caught in large numbers in various countries. Nevertheless, efforts to culture it are underway.

P. (Ma.) japonicus: FAO name, kuruma prawn, similar to the Japanese name, *kurumaebi*. Various other names are used in different regions, e.g. tiger, flower, banded, etc. Widely distributed from Japan to India. Reportedly, it has migrated from the Red Sea to the Mediterranean Sea through the Suez Canal and has been caught by trawlers offshore of several countries. In Japan and the Philippines, commerce relies totally on cultured production.

P. (Me.) kerathurus: Range: East Atlantic Ocean and the Mediterranean Sea. Body length up to 22 cm. Unavoidable in coastal fishing operations of various countries in regions of its distribution. However, numbers not large along the western coast of Africa. FAO name, Caramote prawn.

P. (Me.) latisulcatus: Japanese name *futomizoebi*; FAO name, western king prawn. Widely distributed from the Indian Ocean to the Red Sea. Holds second place to *kuruma* shrimp in Japan and throughout its distribution range.

P. (Me.) longistylus: Restricted to the South China Sea and northeastern waters of Australia. Dwells on sand-mud bottom at depths of 35-55 m. Body length about 15 cm. Because of its habitat, not an important species in fishing operations. FAO name, redspot king prawn.

P. (Me.) marginatus: Japanese name *teraokurumaebi*; FAO name, Aloha prawn. Widely distributed from Japan to the Indian Ocean. Considered a biological resource in Hawaii.

P. (Fe.) merguiensis: FAO name, banana prawn; Japanese name, *bananaebi* or *tenjikuebi*. Large prawn, 24 cm in length. Distributed from the Philippines to the Persian Gulf. Annual catch exceeds 50,000 tons but nonetheless it is also cultured. Often coexists with *P. indicus*.

P. (P.) monodon: Japanese name *ushiebi*; FAO name, giant tiger prawn. Large important species, widely distributed in the West Pacific Ocean. Many local names, e.g. jumbo tiger, black tiger, etc. Body length 33 cm and weight 130 g. Also cultured in many countries.

P. (Fa.) notialis: East and West Atlantic Ocean. FAO name, southern pink shrimp. Abundant at depths of 3-50 m. An important species from the West Atlantic Ocean, especially the Antilles archipelago to Brazil.

P. (L.) occidentalis: Range: Mexico to Peru, in particular Panama, Colombia, and Ecuador. FAO name, western white shrimp.

P. (Fa.) paulensis: Lives at depths of 35-55 m from Brazil to Argentina and until 1967, when it was recorded as a new species, confused with *P. brasiliensis* or *P. aztecus*. FAO name, Sao Paulo shrimp.

P. (Fe.) penicillatus: FAO name, redtail prawn. Range: Taiwan to Pakistan. Important species in Pakistan. Poor record in India due to confusion with *P. indicus* during fishing operations.

P. (Me.) plebejus: Distributed between Queensland in eastern Australia and Victoria in the south. FAO name, eastern king prawn. Large, 30 cm in body length. In Queensland, constitutes more than half the *kuruma* shrimp catch.

P. (L.) schmitti: Important species, ranging from the Caribbean to Brazil. Confused in the past with *P. setiferus* found in the northern part of the Gulf of Mexico. Abundant up to a depth of 30 m; juvenile forms develop near estuaries. FAO name, southern white shrimp.

P. (P.) semisulcatus: Japanese name *kumaebi* or *ashiaka*; FAO name, green tiger prawn. Widely distributed from Japan to the Indian Ocean and has even spread to the Mediterranean Sea through the Red Sea and Suez Canal. Also cultured.

P. (L.) setiferus: FAO name, northern white shrimp. Important species from North Carolina to Gulf of Mexico. In the USA alone, the average annual catch is reportedly 27,000 tons.

P. (Fe.) silasi (Fig 1.6): Reported in 1977 from a depth of 36 m in waters off North Borneo. Body Length about 20 cm. Morphologically resembles *P. merguiensis* and *P. indicus*. Large numbers reported in the past but perhaps confused with closely allied species.

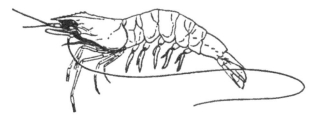

Fig. 1.6. *Penaeus silasi* (Muthu and Motoh, 1979).

P. (L.) stylirostris: Range: Gulf of California to Mexico. In some regions competes for dominance with *P. occidentalis*. FAO name, blue shrimp.

P. (Fa.) subtilis: Range: Antilles archipelago to Brazil. FAO name, southern brown shrimp.

P. (L.) vannamei: Range: Mexico to northern Peru. Particularly abundant in southern part of Mexico, Guatemala, and Ell Salvador. FAO name, whiteleg shrimp.

1.2.2 Small Shrimp Species (*Koebi*) (Caridae)

Compared to shrimps of the genus *Penaeus* (or *Kurumaebi* species), Caridae species are generally smaller. Their value as a biological resource is low but their population high. Their use as processed food is being recognized. The number of species is large and their morphology and ecology varied. So too are their habitats—from fresh water to the deep sea, and they may be pelagic or bottom dwellers.

Seven families are described here: Oplophoridae, Nematocarcinidae, Atyidae, Pasiphaeidae, Palaemonidae, Pandalidae, and Crangonidae.

(1) Family Oplophoridae

These are free-swimming shrimp distributed from the surface layer of the ocean to the middle. Seven genera and 45 species are included in the family. The most common representative is *Acanthephyra purpurea*, which has a strongly flexed body 10 cm in length. During the day members gather at depths of 900-1,800 m and at night ascend to a depth of 200 m. The body is vivid pink and all the legs have luminescent organs. There are 7 species in Japan but only 2 are considered important in fisheries. These are *A. purpurea* and *A. eximia*. The carapace of the former is somewhat circular while in the latter there is a sharp ridge along the middorsal line. Both are widely distributed in Japanese waters, the Indian Ocean, and the Atlantic. The catch is often rich. However, since they are deepsea crustaceans, the flesh content is poor and they store fat; their value as a biological resource is low.

(2) Family Nematocarcinidae

The family comprises only one genus, *Nematocarcinus*, with about 20 species. Most have a benthic habit. The body length is 10-12 cm. They are sometimes caught in large numbers. Even in the experimental offshore operations carried out at Surinam by the Marine Fisheries Resource Development Center, rich collections were made. They were termed SDS (small deep-sea shrimp) and packed into consignments. They were identified as *N. rotundus*. The body breaks easily into cephalothorax and abdomen. In the catalogue prepared by FAO, the species *N. africanus* is said to live at depths of 200-700 m in the East Atlantic Ocean. Its value is low. All 20 species have very long thoracic legs and hence the appellation "spider shrimp". There are 4 species in Japan.

(3) Family Atyidae

Freshwater shrimps include members of this family and members of the family Palaemonidae discussed later. Numa shrimps, as they are popularly called, live in ponds, lakes, and rivers. The pincers of the 1st and 2nd thoracic legs are characteristic.

Numa shrimp *Paratya compressa* and Nuka shrimp *P. c. improvisa* are highly valued as fishing bait. They have little potential as a food resource. Camacuto shrimp *Atya scabra*, ranging from Mexico to Brazil, is about 10 cm long and is used as food in various regions. A few reports indicate the use of other shrimps as food but none has a particularly distinct taste.

(4) Family Pasiphaeidae

Members of this family can be identified by the presence of fine denticles along the cutting margin of the pincers of the anterior 2 pairs of thoracic legs. Many species are free swimming between shallow and deep waters. At present, 7 genera and about 60 species are known. However, only the following 3 species are of interest in the field of fisheries. Even these are not very important.

Genus *Glyphus*: Recently identified with a single species, *G. marsupialis* (Fig. 1.7). *Sympasiphaea annectens* and *S. imperialis*, considered different genera and different species in the past, are now synonyms of *G. marsupialis*. It is distributed in the West Pacific Ocean, Indian Ocean, and the Atlantic. Distribution is poor in the Atlantic Ocean. It is a large shrimp, the body exceeding 20 cm. The cephalothoracic carapace is strongly flexed and has a dorsal ridge. The abdomen of the female is swollen and therefore it is aptly called the kangaroo shrimp. Given the fact that it is a deep-sea species with high fat content and low flesh content, it cannot be considered an effective biological resource.

Fig. 1.7. *Glyphus marsupialis.*

Genus *Leptochela*: Small shrimps, the body length barely reaching 5 cm. Live, they are highly transparent and hence called glass shrimps. There are 12 species and all dwell at the bottom of shallow seas and constitute a major food item for benthic fishes. Only *L. gracilis* has commercial value in fisheries. In China and Korea it is used as food.

Genus *Pasiphaea*: The Japanese species *P. japonica* is called the Japanese glass shrimp. The trade name in Japan is *Bekkoebi*. It is a substitute for *Sakura* shrimp. Since 1953, 200-300 tons have been caught annually. White glass shrimp, *P. sivado*, and pink glass shrimp, *P. multidentata*, are distributed from the northeast Atlantic Ocean to the Mediterranean Sea. Their value is low in fisheries but in Spain and Italy a partial operation is carried out.

(5) Family Palaemonidae

Members of Palaemonidae have pincers in the 1st and 2nd thoracic legs and the 2nd thoracic legs are very long. Mainly members of the freshwater subfamily Palaemoneae are valued in fisheries. Local preferences are well marked in fishing operations. The culture of various types was recently begun and will be discussed in later chapters.

Genus *Macrobrachium*: FAO has recorded 49 species but as of 1952, 70 had been reported and more were added later. The giant river prawn *M. rosenbergii* with a high economic rating is the most well-known species but many others are also used as food in various regions. The number of prawns, large or small, greatly influences the importance of the type. It is felt that marketing will become convenient only when they are caught in sufficient numbers together. Most of the prawns are 10 cm in body length. For purposes of culture only those which grow to more than 15 cm are used. There are 10 species in Japan, of which the most desired species is *M. lar*, i.e., the monkey river prawn.

Genus *Palaemon*: Found in large numbers in the intertidal zones and lakes. Usually small, 5 cm in length, and hence unless caught in large numbers, the market value is negligible. Japan has reported 10 species, of which the most important is the lake prawn *P. paucidens*, which is found in large numbers in Lake Biwa. Besides being an important bait material, it is also used in the preparation of *tsukudani*.

The Baltic prawn *P. adspersus*, rockpool prawn *P. elegans*, and common prawn *P. serratus* of the East Atlantic Ocean have long been known. They have many local names. There are several reports relating to their catching operations and they have the same commercial value as *P. paucidens*.

(6) Family Pandalidae

Members of this family are found in deep waters of cold seas. They are relatively large, very fleshy, and caught in large numbers. Hence they have a high commercial value, second only to *kuruma* shrimp. It is also well known that the males are capable of changing sex, a feature that has drawn much attention among biologists.

Genus *Pandalus*: Japanese waters contain popular species. Of these, 4 have a high production rate: the northern shrimp *P. borealis*, coonstripe shrimp *P. hypsinotus*, Hokkai shrimp *P. kessleri*, and botan shrimp *P. nipponensis*.

The northern shrimp has various local names in Japan, namely, *bokkokuebi, akaebi, tongarashiebi*, and recently the better known term *amaebi*. It is widely distributed in the North Pacific Ocean and the North Atlantic. In other words, it is an extreme northern species. It is caught in large numbers in various regions. The local names are numerous, such as pink shrimp, deepwater shrimp, northern shrimp, etc. It constitutes 80-90% of the shrimp catch in the Bering Sea and Gulf of Alaska. Obviously, it is a very important species. An allied species, the ocean shrimp *P. jordani*, is caught in the waters from Alaska to California.

The coonstripe shrimp *P. hypsinotus* is called *toyamaebi* since it is found in large numbers in Toyama Bay. It has several local names in various regions from the Sea of Japan to the coastal areas of Hokkaido. It is well represented in the Sea of Japan but not so in the Bering Sea and along the coasts of Canada. Even by including the allied species *P. goniurus* (humpy shrimp), it constitutes less than 5% of the total shrimp catch. Incidentally, *P. goniurus* exists in large numbers in the Sea of Okhotsk and its potential as a commercial catch is being considered.

Genus *Pandalopsis*: The Morotoge shrimp *P. japonica* is caught in the northwest Pacific Ocean while the sidestripe shrimp *P. disper* is caught in the northeast Pacific Ocean, but compared to the genus *Pandalus* the importance of this one is low. *Pandalus* purportedly constitutes 5-15% of the shrimp catch in the Bering Sea and Gulf of Alaska.

Genus *Parapandalus*: The Oriental narwhal shrimp *P. spinipes* distributed from Japan to the Red Sea and the narwhal shrimp *P. narval* of the East Atlantic Ocean and the Mediterranean Sea are indicated as potential resources.

Genus *Plesionika*: The golden shrimp *P. martia*, widely distributed in the seas of the world, is found in the local market. However, it is not a common commodity.

Genus *Heterocarpus*: Experimental fishing operations are underway for the armed nylon shrimp *H. ensifer*, which is widely distributed in the world seas. It appears to be a potential resource. According to a survey of the western part of the Indian Ocean, it is caught in considerable numbers at depths of 250-650 m. The catch of the smooth nylon shrimp *H. laevigatus* at depths of 550-800 m in the same waters is also reported to be excellent.

(7) Family Crangonidae

The cephalothorax is flexed dorsoventally and the posterior part of the abdomen is also flexed. The most distinctive feature is the incomplete form of pincers in the 1st pair of thoracic legs. Usually the body length is within 10 cm. These species occur in large numbers from shallow to deep water and several are caught worldwide in large quantities.

Genus *Crangon*: The Japanese sand shrimp *C. affinis* caught in the inner bays of Japan is about 6 cm long and therefore small in size. Nonetheless, its importance in fisheries has often been pointed out. In the northeast Pacific Ocean, Alaska shrimp *C. alaskensis* and cray shrimp *C. communis*, and in the East Pacific Ocean, California shrimp *C. franciscorum*, blacktail shrimp *C. nigricauda*, and bay shrimp *C. nigromaculata* are important species. The 3 species of the East Pacific Ocean once constituted 75% of the shrimp catch of California. Currently, *Pandalus* holds first position.

In the northeast Atlantic Ocean sand shrimp *C. septemspinosa* and in the East Atlantic Ocean, common shrimp *C. crangon* are well known. Japanese fisheries evidence little interest in species of *Crangon*.

Genus *Argis*: *Kuro* shrimp *A. lar* is distributed from the Sea of Japan to the Bering Sea. Another species, *A. dentata* of the northern part of the Sea of Japan, is rich in flesh. The former exists in large numbers in shallow waters. In the seas off Yamagata prefecture, *A. lar* is found at depths of 250-270 m and *A. dentata* at 300 m. In *A. lar*, the dorsal protuberance of the cephalothorax disappears near the posterior edge but in *A. dentata* continues sharply. Even after removal from the water, *A. lar* continues to live for a considerable time and hence can be readily preserved.

1.2.3 Nephrops

Reptantian decapods are broadly divided into Cambaroides and Panulirus groups. Cambaroides includes a few freshwater families and 2 marine families, Nephropidae and *Osate-ebi*. Panulirus is of paramount importance in fisheries and so are some species of lobsters (Nephropidae).

Genus *Nephrops*: The Japanese lobster *Metanephrops japonicus* is caught between winter and spring around Bōso peninsula and the estuary of Sagami River. It lives in the sand-mud bottom at depths of 200-400 m and the 1st pair of thoracic legs is remarkably large and long. The red-banded lobster *M. thomsoni*, distributed from the southern part of Japan to the South China Sea, is used as food in Korea and Taiwan. The Norwegian (or king) lobster *Nephrops norvegicus* is an important species distributed in the East Atlantic Ocean and the Mediterranean Sea. The total catch in 1971 in the countries around the Mediterranean was 2,300 tons. The Caribbean lobster *M. binghami* and others are yet to be developed and the results of experiments are awaited.

Lobsters: At present there are 3 species, of which 2 are very important. These are the American lobster *Homarus americanus* of the northwest Atlantic Ocean and the European lobster *H. gammarus* of the northeast Atlantic Ocean. The average annual catch of the former is 30,000 tons and of the later 2,000 tons. The reasons for the difference in catches of the two are not clear. However, certain conditions no doubt play a role. These are the low temperature of the European waters, the dispersal of larvae away from their habitats due to water current, the stronger larvae of American lobsters, and the ecological competition from other species.

The two species are quite large; American lobsters are unusually large, with a body length of even 1.2 m recorded. The 1st pair of legs have massive pincers (chelae) and the 2nd and 3rd small pincers. Unlike Panulirus they live in holes they dig under rocks. There is much competition with other lobsters for food. The restricted distribution of lobsters in the Atlantic Ocean appears to be related to the history of the earth. The possibility of successfully translocating them to the northeast Pacific Ocean side where there is no competition from Panulirus looks hopeful from the experiments conducted thus far.

1.2.4 Spiny Lobsters

Of the 5 families of Panulirus, only 2 are of value in fisheries, namely Palinuridae and Scyllaridae.

Spiny lobsters: Japanese spiny lobsters of the genus *Panulirus* include 8 species: *P. homarus, P. penicillatus, P. ornatus, P. versicolor, P. japonicus, P. longipes, P. polyphagus,* and *P. stimpsoni*. The range of the most important species, *P. japonicus*, is very narrow while the other species are widely distributed up to the Indian Ocean. *P. ornatus* and *P. polyphagus* are also quite important. In the West Pacific Ocean there are some species such as *P. cygnus* of Australia which are also fairly important. *P. gracilis*, distributed in the East Pacific Ocean from California to Peru, *P. argus* (Fig. 1.8) in the northern part of the West Atlantic Ocean, and *P. laevicauda* of the southern part of the Atlantic Ocean are well represented. Of these, *P. argus* living in the Gulf of Mexico and the Caribbean Sea is found in large numbers. Among the spiny lobsters, *P. argus* is the most important species.

The European spiny lobsters of the genus *Palinurus*, which is identical to the genus *Panulirus*, includes the European species *P. elephas* (Fig. 1.8) and a few species from the waters of West Africa and South Africa. In the case of *Panulirus*, the flagellar part of the antennule is long while in *Palinurus* it is short. Moreover, it has vocal organs in the basal part of the 2nd antenna that can squeak.

In the genus *Jasus*, the flagellar part of the antennule is short and the 2nd antenna has no vocal organs in its basal part. This genus includes

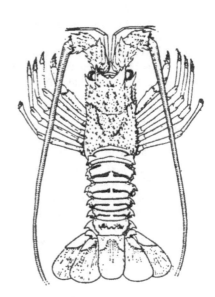

Fig. 1.8: *Panulirus argus* (left) and *Palinurus elephas* (right). (Figure on right from FAO Species Identification Sheets, vol. 2).

7 species called rock lobsters. Since the upper plate of the eye is well developed in *Jasus*, it can be readily differentiated from the real *Panulirus*.

Scyllarides (slipper lobsters): Species of the genus *Scyllarides* are large but not numerous. They live in rocky places and do not come under the usual commercial fishing. *Parribacus japonicus*, like spiny lobsters, also lives under rocks at depths of 10-30 m. It is not sufficiently abundant to warrant commercial fishing. On the other hand, the genus *Ibacus* includes such members as *I. ciliatus* and *I. novemdentatus*, with a dorsoventrally flexed body. Since they live in sand at a shallow level they are often caught.

LITERATURE

The following literature is essential for describing the useful prawns and shrimps of the world.

Holthuis LB. 1980. FAO Species Catalogue, vol. 1. Shrimps and Prawns of the World. An annotated catalogue of species of interest to fisheries. FIR/Sa25(1): i-xvii, 1-271.

As the title suggests, this report deals only with the swimming forms of prawns and shrimps. In addition to recording the FAO names, regional distribution, ecology, size, and potential use in fishery operations are briefly described for each species. References are listed at the end of the volume. The literature referred to in this chapter is drawn from this list. Constraint of space has precluded including all the references so only those extensively used are listed below.

Barnard KH. 1950. Descriptive catalogue of South African decapod Crustacea. Ann. S. Afr. Mus., **38**: 1-837.

Bate CS. 1883. Report on the Crustacea Macrura collected by H.M.S. Challenger during the years 1873-76. Rep. Sci. Voy. H.M.S. Challenger (Zool.) **24**: i-xc, 1-942, pls. 1-150.

Burkenroad MD. 1934. Littoral Penaeidea chiefly from the Bingham Oceanographic Collection, with a revision of *Penaeopsis* and descriptions of two new genera and eleven new American species. Bull. Bingham Oceanogr. Coll., **4(7)**: 1-109.

Chace FA, Jr. 1972. The shrimps of the Smithsonian-Bredin Caribbean Expeditions with a summary of the West Indian shallow-water species (Crustacea: Decapoda: Natantia). Smiths. Contr. Zool., **98**: 1-179.

Crosnier A., Forest J. 1973. *Les crevettes profondes de l'Atlantique oriental tropical*. Faune Trop., **19**: 1-409.

Dall W. 1957. A revision of the Australian species of Penaeinae (Crustacea: Decapoda: Penaeidae). Aust. J. Mar. Freshw. Res., **8**: 136-230.

Holthuis LB. 1950. Subfamily Palaemoninae. The Palaemonidae collected by the Siboga and Snellius Expeditions with remarks on other species. 1. The Decapoda of the Siboga Expedition, Part 10. Siboga Exped. Monogr., **39a(9)**: 1-268.

Holthuis LB. 1959. The Crustacea Decapoda of Suriname (Dutch Guiana). Zool. Verh., **44**: 1-296.

Holthuis LB, Gottlieb E. 1958. An annotated list of the decapod Crustacea of the Mediterranean coast of Israel, with an appendix listing the Decapoda of the eastern Mediterranean. Bull. Res. Counc. Israel, **7B**: 1-126.

Kubo I. 1949. Studies on penaeids of Japan and its adjacent waters. J. Tokyo Coll. Fish., **36**: 1-467.

Man JG de. 1911. Family Penaeidae. The Decapoda of the Siboga expedition, pt. 1. Siboga Exped. Monogr., **39a**: 1-131.

Man JG de. 1920. Families Pasiphaeidae, Stylodactylidae, Hoplophoridae, Nematocarcinidae, Thalassocaridae, Pandalidae, Psalidopsidae, Gnathophyllidae, Processidae, Glyphocrangonidae and Crangonidae. The Decapoda of the Siboga Expedition, pt. 4. Siboga Exped. Monogr., **39a(3)**: 1-318.

Pérez Farfante I. 1969. Western Atlantic shrimps of the genus *Penaeus*. Fish. Bull., **67**: 461-591.

Pérez Farfante I. 1977. American solenocerid shrimps of the genera *Hymenopenaeus*, *Haliporoides*, *Pleoticus*, *Hadropenaeus* new genus, and *Mesopenaeus* new genus. Fish Bull., **75**: 261-346.

Racek AA, Dall W. 1965. Littoral Penaeinae (Crustacea: Decapoda) from northern Australia, New Guinea and adjacent waters. Verh. K. Nederl. Akad. Wet., Natuurk., **56(3)**: 1-119, pls. 1-13.

Schmitt WL. 1921. The marine decapod Crustacea of California with special reference to the decapod Crustacea collected by the United States Bureau of Fisheries steamer "Albatross" in connection with the biological survey of San Francisco Bay during the years 1912-1913. Univ. Calif. Publ. Zool., **23**: 1-359.

Williams AB. 1965. Marine decapod crustaceans of the Carolinas. Fish. Bull. Fish Wildl. Serv., **65**: 1-298.

Besides the above important references, literature concerning the reptantian shrimps is listed below.

Burkenroad MD. 1981. The higher taxonomy and evolution of Decapoda (Crustacea). Trans. San Diego Soc. Nat. Hist., **19**: 251-268.

Chace FA Jr., Dumont WH. 1949. Spiny lobsters—identification, world distribution, and U.S. trade. Comm. Fish. Rev., **11(5)**: 1-12.

Chace FA Jr., Manning RB. 1972. Two new caridean shrimps, one representing a new family, from marine pools on Ascension Island (Crustacea: Decapoda: Natantia). Smiths. Contrl. Zool., **131**: 1-18.

Chace FA Jr., Brown DE. 1978. A new polychelate shrimp from the Great Barrier Reef of Australia and its bearing on the family Bresiliidae (Crustacea: Decapoda: Caridea). Proc. Biol. Soc. Wash., **91**: 756-766.

Fischer W. (ed.). 1978. Crustaceans. Mediterranean and Black Sea (Fishing Area 37). FAO species identification sheets for fishery purposes, vol. 2.

Fischer W. (ed.). 1978. Lobsters, shrimps and prawns. Western Central Atlantic (Fishing Area 31). FAO species identification sheets for fishery purposes, vol. 6.

George RW. 1972. South Pacific Islands—Rock lobster resources. A report prepared for the South Pacific Islands Fisheries Development Agency, FI: RAS/69/102/9: iii + 42.

George RW., Holthuis LB. 1965. A revision of the Indo-West Pacific spiny lobsters of the *Panulirus japonicus* group. Zool. Verh., **72**: 2-36, pls. 1-5.

George RW., Kensler CB. 1970. Recognition of marine spiny lobsters of the *Jasus lalandii* group (Crustacea: Decapoda: Pallinuridae). N.Z.J. Mar. Freshw. Res., **4**: 291-311.

Glaessner MF. 1960. The fossil decapod Crustacea of New Zealand and the evolution of the Order Decapoda. N.Z. Geol. Surv., Palaeont. Bull., **31**: 1-79.

Glaessner MF. 1969. Decapoda. In: R.C. Moore (ed.). Treatise on Invertebrate Paleontology, Part R, Arthropoda 4, vol. 2, pp. R399-R566.

Gruvel A. 1912. *Contribution à l'étude générale systématique et économique des Palinuridae.* Ann. Inst. Océanogr., **3(4)**: 5-56, pls. 1-6.

Hayashi K. 1981-83. Classification and Ecology of Japanese Shrimps, 1-10. Ocean and Organisms, **3**: 368-371, 452-455; **4**: 46-49, 110-113, 188-191, 292-295, 354-357, 438-442; **5**: 32-35, 102-104.

Note: The following references pertain mainly to the family Penaeidae, although other families are also discussed.

Holthuis LB. 1947. Biological results of the Snellius Expedition. XIV. The Decapoda Macrura of the Snellius Expedition. I. The Stenopodidae, Nephropsidae, Scyllaridae and Palinuridae. Temminckia, **7**: 1-178, pls 1-11.

Jensen AJC. 1967. The Norway lobster, *Nephrops norvegicus*, in the North Sea, Skagerak and Kattegat. Proc. Symp. Crust., Ernakulam, **4**: 1320-1327.

Kensley BF. 1968. Deep sea decapod Crustacean from west of Cape Point, South Africa. Ann. S. Afr. Mus., **50**: 283-323.

Kensley BF. 1977. The South African Museum's Meiring Naude cruises. Part 5. Crustacea, Decapoda, Reptantia and Natantia. Ann. S. Afr. Mus., **74**: 13-44.

Kim HS. 1977. Macrura. Illustrated Flora and Fauna of Korea, **19**: 1-414.

Lee DA., Yu HP. 1977. The penaeid shrimps of Taiwan. JCRR Fish. Ser., **27**: 1-110.

Mauchline J., Murano M. 1997. World list of the Mysidacea, Crustacea. J. Tokyo Univ. Fish., **64**: 39-88.

McLaughlin RA. 1980. Comparative Morphology of Recent Crustacean. W.H. Freeman and Co., San Francisco, 177 pp.

Muthu MS., Motoh H. 1979. On a new species of *Penaeus* (Crustacea, Decapoda: Penaeidae) from North Borneo. Res. Crust., **9**: 64-70.

Silas EG. 1967. On the taxonomy, biology and fishery of the spiny lobster *Jasus lalandei frontalis* (H. Milne-Edwards) from St. Paul and New Amsterdam Islands in the southern Indian Ocean, with an annotated bibliography on species of the genus *Jasus* Parker. Proc. Symp. Crust., Ernakulam, **4**: 1466-1520.

Tachibana K. 1981. Culture of *Homarus. Kaiyokagaku* (Oceanography Symp.), **132**: 872-884.

Tachibana K., Kajiwara T. 1981. Distribution of *Homarus. Kaiyokagaku* (Oceanography Symp.), **132**: 821-832.

Takeda M., Tadahito S. 1982. Scientific name of kangaroo shrimp. *Kokuritsu Kasen Senpo* (Nat'l. Sci. Rep.), **15**: 181-185.

Tirmizi NM. 1971. *Marsupenaeus*, a new subgenus of *Penaeus*, Fabricius, 1798 (Decapoda, Natantia). Pakistan J. Zool., **3**: 193-194.

2

Ecology of Prawns and Shrimps

2.1 COMMON HABITATS OF PRAWNS AND SHRIMPS

A survey of the natural habitats of animals is difficult due to various factors even in the case of terrestrial organisms. Such a survey of marine creatures is all the more difficult.

About 2,300-2,400 species of prawns and shrimps are said to exist. Even in the case of *Panulirus japonicus (ise-ebi)* and *Penaeus japonicus (kurumaebi)*, closely associated with the life of Japanese people since ancient times, the exact conditions of the habitats for a natural life history are not clearly known. The same can be said for another species of *Penaeus*, i.e. *Penaeus monodon (ushiebi)*, for which the scale of culturing has rapidly increased in Southeast Asia, commencing with Taiwan.

In order to promote culturing, it is of utmost importance to have a complete understanding of the biological characteristics of prawns and shrimps and to utilize this knowledge to obtain an ideal production. Most of the shrimps of the genus *Penaeus* spawn in open seas and the tiny larvae, barely 1/3 mm, grow to a few mm, then migrate to the coastal region where they spend the juvenile stage. Thereafter, they once again migrate to open seas, attain maturity and lay eggs. The repetition of this life cycle has been clearly worked out. Nevertheless, not much is known about the mechanism which initiates this type of behavior nor the rhythmicity of migration. The technique of artificially growing the ovipositing mother shrimps of *Penaeus japonicus* and *Penaeus monodon* cannot be deemed wholly successful. A complete understanding of these mechanisms would pave the way to success.

Most of the research pertaining to the migration and circular movements of species of *Penaeus* have reported a close relationship between migration and seasons and factors associated with migration such as food, water temperature, spawning, and optimum salinity. *Penaeus chinensis* (locally called *taishoebi* or *koraiebi*), which grows to a body length of 27 cm, starts spawning in the coastal regions of Hong Kong and northwestern areas of the Korean peninsula; maximum spawning occurs in July. The young shrimps born and developing in the current year form groups between early autumn and early winter and descend southward to overwinter in the central part of the Yellow Sea and in the deep waters west of Saishu Island. With the advent of spring they ascend northward to reach the spawning grounds (Kasahara, 1948; Ikeda, 1962). *P. setiferus* breeds along the eastern and southern coasts of the USA and moves to offshore waters for the final phase of growth. *P. paulensis* migrates between southern parts of Brazil for breeding in the gulf and offshore waters for advance growth.

To date, 19 species of family Palaemonidae and 17 species of family Atyidae have been reported in Japan. However, all these species are small or medium in size and for commercial purposes only the members of *Macrobrachium* are targeted.

Macrobrachium lar (konjintenagaebi)	15 cm
M. latimanus (kotsunotenagaebi)	12 cm
M. formosense (minamitenagaebi)	12 cm
M. nipponense (tenagaebi)	9 cm
M. japonicum (yamatotenagaebi)	9 cm
M. equidens (subesubetenagaebi)	8 cm

Of these, *M. nipponense* has been caught since ancient times in rivers and lakes of various regions. The catch is particularly large in Kasumigaura, Biwako, and Shinjiko and this species is well known in the respective local regions (Tables 2.1 and 2.2).

Table 2.1: Total annual catch of freshwater shrimp for the past 10 years. Catches in Kasumigaura and Kitaura and the percentage of total catch

Year	Total	Kasumigaura, Kitaura (percentage of total catch)		Year	Total	Kasumigaura, Kitaura (percentage of total catch)	
	Tons	Tons	%		Tons	Tons	%
1971	4,982	4,011	81	1976	5,301	3,250	61
1972	5,272	3,911	74	1977	6,151	4,120	67
1973	6,441	3,383	53	1978	9,276	4,764	51
1974	5,739	3,636	63	1979	7,639	4,119	54
1975	6,840	4,972	73	1980	5,846	3,656	63

Table 2.2: Lakewise, riverwise, and prefecturewise catch of freshwater shrimps in 1980 (4 examples are given for each category)

Lakes		Rivers		Prefectures	
	Tons		Tons		Tons
Kasumigaura	3,101	Rinekawa	63	Ibaraki	3,695
Shinjiko	1,015	Chikagokawa	23	Shimane	1,017
Biwako	725	Shimantokawa	22	Shiga	726
Kitaura	555	Nagaragawa	12	Chiba	86

When the total catch of freshwater prawns and shrimps is compared to the catch of marine prawns and shrimps, the difference is approximately 10%. Fishing is one of the important marine industries in Kasumigaura, Biwako, and Shinjiko. The freshwater shrimp catch of Japan including these regions cannot be expected to expand. This is because topographically Japan is a narrow and long archipelago extending north-south. Moreover, it has too many mountain ranges with tall mountains. Though the annual rainfall is 1,900 mm, given the steep declination of the rivers, the water flows down rapidly. Moreover, there are very few big rivers and lakes. Further, the weather, especially in northern Japan, is too cold for the survival of freshwater shrimps. These factors not only adversely affect the propagation of freshwater shrimps, but also hinder the development of the entire freshwater fishing industry.

Studies on the ecology of natural habitats of freshwater shrimps are far fewer than those on marine shrimps and, in fact, this aspect is not known for most of the species. In this book, the following aspects are discussed for the shrimps of *Macrobrachium*, based on a several-year survey by experts.

Among the existing freshwater shrimps, some species are wholly such in the sense that they spend their entire life in fresh water, while others spend a temporary period, in particular the larval period, near or in estuaries. On attaining the juvenile stage they gradually migrate inland to lakes and rivers and spend their remaining life in fresh water. Some spend most of their life in estuaries. The latter two types cannot be clearly distinguished as there are some that move back and forth between estuaries and fresh waters. Numa shrimps belong to the first type. Genus *Macrobrachium* belongs to the second type. In certain regions, such as Suwako, cases have been reported of shrimps moving inland and repeating their life history in fresh water only. When prawns were made to spawn indoors (laboratory) and the hatchlings reared in culture media with various concentrations of salinity, the larvae were found to survive and grow well in solutions containing 5-15‰ salinity.

2.2 DISTRIBUTION AND BEHAVIOR OF PRAWNS

In a wide area ranging from the montane region in the western part of Kochi Prefecture to the Shimanto River opening into the Pacific Ocean (Fig. 2.1), three species of prawns, namely, *Macrobrachium nipponense (tenagaebi)*, *M. formosense (minamitenagaebi)*, and *M. japonicum (yamatotenagaebi)*, live in large numbers. The main stream of this river is 185 km long and the area covered 2,270 km^2. According to annual reports on river flow published by the Ministry of Construction (1952-1968), the average annual total flow is 3.8×10^9 m^3. However, this region is in one

Fig. 2.1. Shimanto River catchment basins (• indicates locations surveyed).

of the few heavy rain belts and undergoes seasonal variations with a maximum flow rate of 13,880 m³ s⁻¹ during the typhoon period in autumn (the region is typhoon prone), minimum of 11.8 m³ s⁻¹ during the drought period, and 51 m³ s⁻¹ during normal flow. Moreover, as shown in Fig. 2.2, there is hardly any difference in height above sea level up to a distance of 13 km along the upstream from the estuary. The height above sea level at Ekawazaki at a distance of 51 km upstream from the estuary is 40 m and at Taiseimachi at a distance of 100 km upstream, 140 m. The inclination of the moderate and low current regions is relatively gentle. In the main stream there are no dikes, dams or other artificial structures. Except for Nakamurashi, there is no other town or factory along the entire length of the river. It is indeed a rare example of a major river left free from human disturbances to the natural environment. Other than fishing, the river is hardly used.

In the low current region shrimp fishing is carried out by two methods: brush trap and stone piling. In the moderate current region, juvenile shrimps are caught by ascending trapnet between early and late summer and adult shrimps by descending trapnet.

The distribution of habitats of adult shrimps is shown in Fig. 2.2. It can be seen from the Figure that *M. formosense* and *M. japonicum* are widely

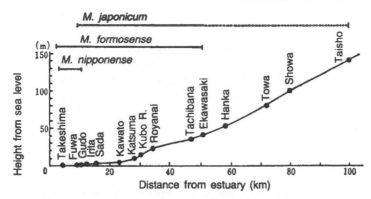

Fig. 2.2. Distribution of habitats of three species of *Macrobrachium* and inclination of Shimanto R. (• indicates locations surveyed).

distributed and intermixed. *M. nipponense* is mainly found in the estuary and coexists to some extent with *M. formosense*.

M. nipponense lives in large numbers in the confluence of the Nakasuji, Takeshima, and Nabeshima rivers which form a common estuary. It is almost absent in the region of water flow. In the main stream, at a place called Sada, 16 km upstream from the river mouth, the percentage of mixing is only 0.05% and above this point negligible. Among the 3 species of *Macrobrachium*, *M.-nipponense* has the lowest resistance to water current; it is a still-water type found in the estuary.

The habitats of *M. formosense* are distributed from the environs of the estuary up to Ekawasaki, 50 km upstream. The sites from which it is caught during the day are either backwaters or deepwater pits formed by the entrance of water into depressions along the river banks. The water in both cases is stagnant and the bottom mud or sand mixed with mud. Should stones or pebbles be present, the shrimps hide in the suspended mud collection or under plants, pieces of wood, pebbles, or stones.

The Ato River, which merges with the Shimanto near the estuary, is a relatively large tributary. At Isaihara, about 18 km from the confluence, a dike of 7 m height has been built and *M. japonicum* and *M. formosense* preferentially dwell beneath it. Upstream of the dike only *M. japonicum* is found; *M. formosense* is totally absent. Perhaps the dike obstructs the upward migration of *M. formosense*.

M. japonicum does not live near the river mouth but in the region of the lowest stream near Fuwa, about 8 km upstream from the estuary. Upstream it is widely distributed about 100 km from the river mouth. However, upstream beyond Ekawasaki the prawn population is relatively less than downstream. Between Fuwa and Ekawasaki it coexists with *M. formosense*. Thus the range of habitats in which *M. japonicum* coexists with *M. formosense* is wide but it also lives in a considerably extended upstream area.

The environment of the habitats occupied by *M. japonicum* differs notably from that of the other 2 species. The most striking difference is that *M. japonicum* avoids still waters and prefers shoals where the water flow is steady. The velocity of flow should be 20-50 cm s^{-1} and the nature of the bottom such that the prawns can hide under pebbles or small stones.

All these habitats are occupied during the day; *M. japonicum* prowls widely and may even appear in places dominated by *M. formosense* during the day. Contrarily, *M. formosense* stays put in shoals both day and night.

Of the 3 species of prawns discussed above, *M. japonicum* has the best developed thoracic legs and the greatest strength against water current. Concomitantly, it better maintains activity in air and while creeping on land; it can climb a dike or wall of a few meters height at places still wet from night dew. This behavior is observed not only among adult prawns, but also among the juvenile forms. This appears to be one of the reasons for the extremely wide range of distribution of habitats of *M. japonicum* to upstream regions. When this species is placed in ponds or tanks (with still water) along with *M. formosense*, individual *M. japonicum* almost always escape at night by crawling over dikes and walls. This behavior is perhaps an expression of the ecological feature related to living in shoals in the past.

In Shimanto River, *M. japonicum* spawn from June to mid-August in all the upstream, midstream, and downstream regions. The larvae that hatch are carried by the water current to the estuary and further to the open sea. Hardly any study has been carried out on the distribution, growth, migration (downstream, upstream) and other aspects of this species and future studies are eagerly awaited. At a constant point in the estuary, the author and his colleagues collected water (using a pump) at every 1 m depth down to 8 m (water depth 9 m). Collections were made every hour continuously for 24 hours. In this way, the appearance of larvae was observed in relation to time lapse and vertical distribution. The results are shown in Figs. 2.3 and 2.4. Before sunset, around 6:00 p.m. when it is still bright, the larvae are confined to a 6-m depth. At 7:00 p.m., when dusk is setting in, the larvae begin to appear at the upper layer; at 8:00 p.m., when it is completely dark, they are seen only in the upper layer and remain there until morning, i.e., until it becomes light. Around 6:00 a.m. they once again appear only in the lower layer. Between 9:00 a.m. and 3:00 p.m. when brightness (radiance) is maximum, the larvae are seen nowhere from the surface to the lower layer. The eggs hatch after sunset and the time when the larvae hatched in various regions migrate to the river mouth differs depending on the distance to the estuary. The

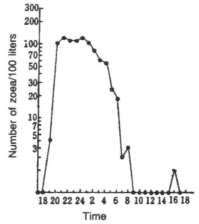

Fig. 2.3. Timewise appearance of zoea.

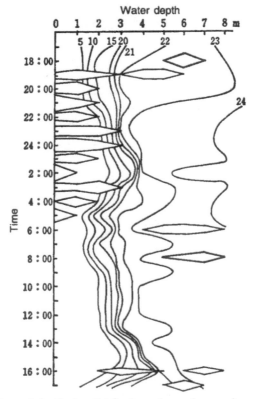

Fig. 2.4. Timewise and depthwise distribution of zoea (curves show chlorinity expressed by specific gravity).

survey could not confirm a constant pattern in time of appearance of the larvae in the estuary. At the same time, the temperature and specific gravity were measured, in particular the density. Since the desirable concentration of chlorine (5-10‰ Cl) is known from laboratory larval cultures, they were expected to be distributed in large numbers at a particular specific gravity (density) layer. However, as shown in Figure 2.4, larval distribution is not at all related to chlorine concentration but closely related to brightness.

When the changes in the catch during the course of a day using an ascending trapnet at Kawato, about 23 km upstream from the river mouth, were studied (Table 2.3), it was found that during the day only gobies were caught and not a single prawn. But as brightness declined to a few lux after sunset, the gobies disappeared in the catch and prawns of *M. formosense* and *M. nipponense* were found instead. This type of survey was conducted several times and it was always observed that the catch reached maximum 2-3 hours after sunset and thereafter gradually decreased up to early dawn, becoming nil around 7:00 a.m. when the sky had brightened.

Table 2.3: Timewise catch of juvenile prawns in ascending tr. met during survey at Kawato (Ohno et al., 1977)

Date	Time	Bright-ness	Water temp (°C)	Number			Total weight (g)		
				M. nippo-nense	*M. for-mosense*	Total	*M. nippo-nense*	*M. for-mosense*	Total
July 16	18 : 00	2800	24.5	1	0	1	3	0	3
	19 : 00	600	24.5	0	0	0	0	0	0
	20 : 00	0	24.3	19	187	206	52	85	137
	21 : 00	0	24.3	92	444	536	98	229	327
	22 : 00	0	23.7	86	915	1001	54	465	519
	23 : 00	0	23.8	29	328	357	26	145	171
	24 : 00	0	23.6	67	249	316	53	120	173
July 17	1 : 00	0	23.6	86	138	224	70	70	140
	2 : 00	0	23.5	95	150	245	56	73	129
	3 : 00	0	23.5	56	153	209	42	80	122
	4 : 00	0	23.5	63	173	236	45	80	125
	5 : 00	140	23.5	63	58	121	62	27	89
	6 : 00	10,000	23.5	10	2	12	4	4	8
	7 : 00		23.6	0	4	4	0	2	2

These surveys support the earlier view that both *M. formosense* and *M. nipponense* are nocturnal but the intensity of their activities cannot always be categorized into 2 peaks, e.g. "soon after sunset" and "before dawn". *M. formosense*, which lives in backwaters during the day, can be called a typical peak form.

Table 2.4: Important species of prawns of North and South America, their body length, size (diameter) of eggs, and habitats (Holthuis, 1952; and others)

Species	Body length (mm)	Egg size (dia) (mm)	Regional habitats
Macrobrachium americanum	230	0.4–0.7	Western North America
M. amazonicum	150	0.6–0.8	Guiana—Paraguay
M. panamense	134		Honduras—Ecuador
M. inca	105	0.45–0.7	Western North America
M. rathbunae	105	0.5–0.7	Western North America
M. acanthurus	166	0.47–0.65	West Indies—Eastern North America—Brazil
M. tenellum	116	0.5–0.6	California—Northern Peru
M. ohione	104	0.35–0.5	Central North America
M. heterochirus	135	0.35–0.5	Southern North America—Mexico—Brazil
M. carcinus	233	0.44–0.67	West Indies—Eastern North America
M. rosenbergii	350	0.48–0.58	Southeast Asia

Details of the regional habitats and the size of a few species of prawns used as food in various regions of North and South America where the survey of prawn species is in progress are given in Table 2.4. Note: *M. rosenbergii* is indigenous to Southeast Asia.

2.3 SPAWNING PERIOD

The spawning period of prawns and shrimps of Japan varies even for the same species depending on the region. A comparison of the spawning periods at Matsushima Bay, Mikawa Bay, Setonaikai, and Ariakekai indicated that this variation may range from 15–45 days. Generally it ranges between the latter half of spring and midautumn. The spawning period of various species of prawns and shrimps has been studied and reported by several experts, e.g. Ohta (1949), Kubo (1950), Maekawa *et al.* (1953), Hachiyanagi *et al.* (1954-1957), Yasuda (1957), Sato (1957), Ikematsu (1963), and many others. While Yasuda studied in detail the spawning period of species of prawns and shrimps in Setonaikai, Sato carried out similar studies in Ariakekai. According to both authors, without exception all members of the genus *Penaeus* spawn between late spring and midautumn. *Metapenaeopsis barbata* (*akaebi* in Japanese) is the first to oviposit in Setonaikai, starting around June. It is followed by *Trachypenaeus curvirostris* (*saruebi*) and *Atypopenaeus compressipes* (*maimaiebi*). The latter continues to spawn to the end of October. On the whole, the three species spawn for 2-4 months.

Shrimps of Caridea continue to spawn from early spring to late autumn and spawning lasts for 6-8 months. In these shrimps, when the larvae hatching from eggs attached to the abdominal legs or the lateral plates of the abdominal segments are freed, the mother shrimps undergo molt and within a few days are ready for the next round of spawning. All the species of shrimps of Caridea have short-period generations (early hatching larvae grow, attain maturity, and spawn to produce another generation in the same year). The active period of spawning may display 2 peaks in some and none in others. In species with 2 typical forms the first peak occurs during spawning of the longer generation and the other peak during spawning of the short-term generation. In species in which 2 typical forms are not distinct for the spawning period, the earlier half includes spawning of the longer generation and spawning of the short-term generation accelerates from around the middle and appears in the latter half.

In the case of smaller species of penaeids such as *Acetes japonicus*, *Parapenaeopsis tenella*, and *Trachypenaeus curvirostris*, the early hatching larvae grow and spawn in the same year. That is, they are a short-term generation. *Penaeus japonicus* (in Setonaikai spawning occurs in summer), *Penaeus semisulcatus* and *Metapenaeus ensis*, which are somewhat larger, as well as the smaller shrimps also spawn late. In these species and also in *Atypopenaeus compressipes*, *Metapenaeopsis acclivis*, *Metapenaeopsis dalei*, and *Metapenaeopsis lamellata*, short-term generations are not observed.

Yasuda (1967) has reported on the spawning periods of deep-sea shrimps collected between Kumano Strait and offshore Miyazaki at depths of 200-500 m. Though about 40 species of shrimps were collected, spawning periods were observed only for the following 8:

Pandalus nipponensis (botonebi)
Parahaliporus sibogae (higenagaebi)
Plesionika martia (okinosujiebi)
Heterocarpus sibogae (minoebi)
Crangon sagamiense (sokoebijako)
Parapenaeus lanceolatus ⎱
Metapenaeopsis coniger ⎬ *(togesake-ebi)*
M. lata ⎰

According to Yasuda, most of these species spawn between late autumn and early spring, in particular during the winter months of December-March. Among them, *Pandalus nipponensis* is famous for its ability to change sex. In other words, it is male when young but transforms to female on becoming an adult.

Kubo (1948, 1949, 1950) has reported on the biological and ecological aspects of *Macrobrachium nipponense* of Kasumigaura. The report includes details of growth, maturation, and spawning. In an effort to study the mechanism of adding resources, the author and his colleagues carried out a survey in the latter half of the 1970s. One of the aspects covered was the course of gonadal development on the basis of GSI (Gonosomatic Index). The percentage of incubating shrimps and the appearance of juvenile shrimps were observed to estimate the spawning period. According to this study, the longer period generation, which has overwintered, begins to prepare for spawning from around April. In mid-May shrimps with 5-10 GSI appear and some even bear ripe eggs. However, no incubating shrimps could be collected during this period. By June shrimps of more than 10 GSI as well as spawning and incubating individuals were found. In late June the overall GSI increased and the percentage of incubating shrimps was 20. Spawning in the real sense began only in July. The percentage of incubating shrimps increased to more than 80. The eggs were in the post-eyespot stage and close to hatching. The GSI was 10 or more. Spawning → incubating → rematuration of gonads progressed simultaneously. This phenomenon continued to mid-August. Thus several spells of spawning occur in one spawning period.

2.4 GONADAL DEVELOPMENT

In all species of shrimps and prawns, gonads are present in the cephalothoracic part below the heart and above the liver. In *Penaeus japonicus*, *Penaeus monodon*, and *Penaeus orientalis* (also known as *P. chinensis*), the posterior lobes of the paired ovaries grow almost to the posterior end of the abdomen (Fig. 2.5). However, in *M. nipponense* and other small shrimps, the ovaries are confined to the cephalothoracic region and never extend into the abdomen.

While studying the ecology and life history of penaeids, several authors also studied the development of ovaries and maturation of ova: Hudinaga (1942), Thorson (1946), King (1948), Williams (1955), Yasuda (1956), and Oka (1965). The course of maturation was observed in several phases. Ikematsu (1963) studied the shrimps of Ariakekai. The degree of ovary maturation is judged by the color and size of the ova and their morphological condition. Ikematsu categorized the ovaries as follows:

Immature: Including the eggs soon after liberation. They are colorless and semitransparent. When fixed in formalin they turn white.

Slightly mature: As they develop the eggs become pale yellowish-green. When fixed in formalin they turn creamy or yellow.

Mature: Eggs greenish; upon fixation they turn orange or dull pink.

In the "seedling" production of *P. japonicus* and *P. monodon*, it is necessary first of all to identify the mother shrimps with substantial mature eggs. Seedling production of *P. japonicus* has already reached several hundred million seedlings. All the mother shrimps used in seedling production were collected from natural resources. Their degree of maturation was judged by the naked eye—color and size of the ovary (Figs. 2.6 and 2.7).

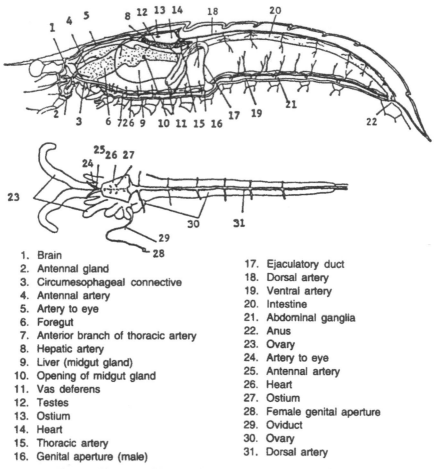

1. Brain
2. Antennal gland
3. Circumesophageal connective
4. Antennal artery
5. Artery to eye
6. Foregut
7. Anterior branch of thoracic artery
8. Hepatic artery
9. Liver (midgut gland)
10. Opening of midgut gland
11. Vas deferens
12. Testes
13. Ostium
14. Heart
15. Thoracic artery
16. Genital aperture (male)

17. Ejaculatory duct
18. Dorsal artery
19. Ventral artery
20. Intestine
21. Abdominal ganglia
22. Anus
23. Ovary
24. Artery to eye
25. Antennal artery
26. Heart
27. Ostium
28. Female genital aperture
29. Oviduct
30. Ovary
31. Dorsal artery

Fig. 2.5. Diagram of *Penaeus japonicus* (Hiroshima Bunridae, 1949).

Among the freshwater shrimps, the author and his colleagues studied *Macrobrachium rosenbergii* for some years. They observed that the ovary in an immature condition is colorless, turning pinkish-orange with maturation. It is clearly visible through the carapace. When the ovary is visible and the integument, originally dark, becomes light orange, this indicates that the shrimp is approaching prespawning molt.

The ratio of weight of a mature ovary to overall body weight varies, according to Ikematsu (1963), from species to species. Most lie in the range of 7-8% to 14-15%. For *M. nipponense* also this ratio is 14-15%.

Reports on the development and maturation of testes are few and mostly deal with the morphology of spermatophores and spermatozoa. Mattews (1954) recorded the formation of spermatozoa and spermatophores in *Enoplomatopus occidentalis*, King (1948) studied the

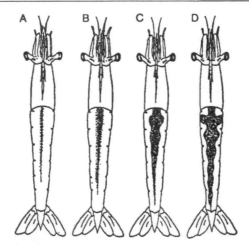

A. Ovary yet to develop   B. Immature ovary
C. Slightly developed ovary   D. Mature ovary

Fig. 2.6. Development of ovary of *Penaeus monodon* (Motoh, 1981).

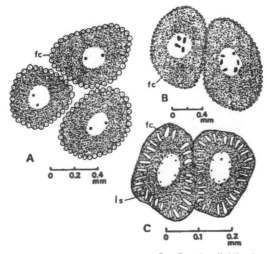

A. Ova of young ovary   B. Ova in slightly developed ovary
C. Slightly mature ova; fc: follicular cells; js: jellylike substance

Fig. 2.7. Cross section of ova of *Penaeus japonicus* (Hudinaga, 1942).

reproductive organs of *Penaeus setiferus*, while Mallex and Bawab (1974) have reported the spermatophore formation of *Penaeus kerathurus*. Mattews (1951), Berry (1970), Berry and Heydorn (1970) and Silberbauer (1971) have reported the structure of the vas deferens and the formation of spermatophores for several spiny lobsters. Some of these reports also discuss the mechanism of fertilization, an aspect that will be discussed later.

Spiny lobsters belong to the group which lacks a seminal vesicle. The male reproductive organs, in particular the vas deferens and the spermatophores of Caridae, belonging to the same group, have hardly been studied to date.

Chow *et al.* (1981) studied *M. rosenbergii* (Fig. 2.8), *M. nipponense, M. formosense, M. lar, M. japonicum, M. latidactylus,* and *Palaemon paucidens* (in part unpublished).

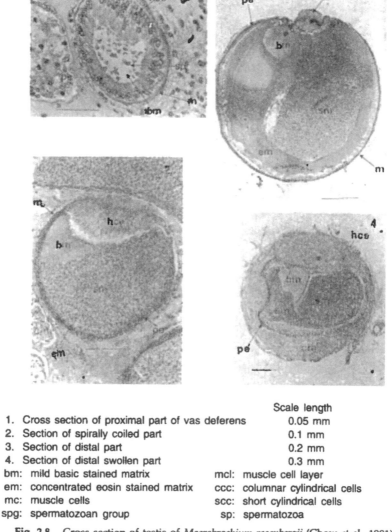

		Scale length
1.	Cross section of proximal part of vas deferens	0.05 mm
2.	Section of spirally coiled part	0.1 mm
3.	Section of distal part	0.2 mm
4.	Section of distal swollen part	0.3 mm

bm:	mild basic stained matrix	mcl:	muscle cell layer
em:	concentrated eosin stained matrix	ccc:	columnar cylindrical cells
mc:	muscle cells	scc:	short cylindrical cells
spg:	spermatozoan group	sp:	spermatozoa

Fig. 2.8. Cross section of testis of *Macrobrachium rosenbergii* (Chow *et al.,* 1981).

The male reproductive organs of all the species of *Macrobrachium* are morphologically very similar. There is a pair of anteroposteriorly elongated testes and a pair of vas deferens. Each vas deferens opens either in the center or in the posterior part of the testis. The spirally coiled part of the testis is followed by the distal slightly swollen part and is composed of contractile smooth muscles. It ends at the male genital aperture (Fig. 2.9). The testes are covered by a thin membrane. These are composed of a large number of minute vesicles and resemble grape clusters.

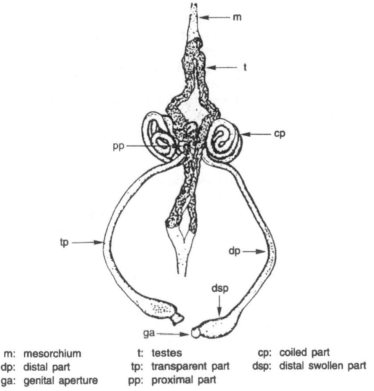

m: mesorchium	t: testes	cp: coiled part
dp: distal part	tp: transparent part	dsp: distal swollen part
ga: genital aperture	pp: proximal part	

Fig. 2.9. Male genital organs of *M. rosenbergii* (Chow *et al.*, 1981).

Figure 2.9 shows the histological nature of the 4 morphologically different regions of the vas deferens of *M. rosenbergii*. The 4 regions are the narrow and short part near the testes, the spirally coiled part, the transparent part near the distal end, and the swollen distal part. At the base, close to the testis, thin muscle fibers form the periphery and cylindrical cells lie inside. The central cavity contains spermatozoa mixed in a gelatinous matrix. In the spirally coiled part, a mild basic stain stainable matrix secreted by large cylindrical cells and concentrated eosin stainable matrix secreted by short cylindrical cells surround the spermatazoan groups. The histological section of the distal part does not differ much from that of the proximal section of the spiralled part. However, the spermatozoa and the two matrices just mentioned are consolidated in this part. The wall of the swollen distal part is composed of a well-developed muscle layer and appears to play an important role in the ejaculation of spermatozoa during copulation.

2.5   ECDYSIS

As mentioned earlier, prawns and shrimps, like other crustaceans, undergo periodic ecdysis during growth. (Insects and roundworms also undergo ecdysis. Even among vertebrates the ecdysis of snakes, falling of feathers in birds, and falling of hair in mammals are akin to ecdysis.) Ecdysis is a phenomenon wherein the hard cuticular layer covering the epidermis is discarded and is essential for ascertaining body growth.

In the case of crustaceans, prior to molt the calcium present in the old cuticle is dissolved in the blood and when the new cuticle has grown, the stored calcium is transferred back. In this way a decrease in calcium is precluded.

Ecdysis is accelerated by the secretion of ecdysone. However, in the case of crustaceans, it is said that the hormone is secreted from the endocrine gland known as the Y-organ (also as the Y-gland) present in the antennal segment or maxillary segment. This is controlled by the X-organ (sinus gland) which secretes the ecdysis-regulating hormone. The X-organ-sinus gland (also known as the hemocoelomic gland) which secretes the ecdysone is present in the eye stalk of shrimps. It is said that the hormone produced by the neurosecretory cells of the X-organ passes through the nerve fibers of the X-organ and gets released from the sinus organ into the hemocoel. The hormone released by the sinus organ is commonly termed the eye-stalk hormone or the sinus-organ hormone. Besides the ecdysis-regulating hormone, the sinus-organ hormone contains other important hormones such as the ovary maturation-regulating hormone and hormones produced by the brain, which include the body color-changing hormone, blood sugar-regulating hormone, heartbeat-increasing hormone, and various hormones to regulate the metabolism of water, carbohydrates, Ca, and P.

Observations on the ecdysis of various types of indoor cultured shrimps indicated that ecdysis can be broadly classified into 3 categories:

Category 1: Ecdysis occurs for the purpose of growth. In the case of penaeids, after hatching the nauplius undergo molt 6 times within 35-40 hours at 25-26°C water temperature and transform into zoea. This is followed by the stages of mysis and the postlarval period. Thus during the larval period molts for growth are repeated accompanying morphological transformations. These molts occur in short intervals. In the case of *Macrobrachium rosenbergii*, the larvae hatch as zoea of the first stage, undergo the first molt within a day or two at 27-28°C water temperature and enter the second stage of zoea. Molts are repeated at intervals of 2-3 days. After 11 molts the larvae metamorphose into the postlarval (juvenile) stage. During this period the body grows from 1.9 mm to about 8 mm. Molts continue even during the postlarval period at intervals of a few days but the intervals are longer than during the younger stages. When larvae were cultured at 27.5 ± 0.5°C water temperature and 0-5.01‰ Cl, molting intervals were observed as follows: 4-7 days for juvenile shrimps 2-3 cm in body length; 9-11 days for body length 4-6 cm; 13-15 days for body length 7-9 cm; 28-35 days for mature female shrimp 13-15 cm long and 34-42 days for mature female shrimp 15-17 cm long. In the case of mature female shrimp, molting is associated more with spawning than with growth.

In the case of crustaceans with a hard cuticular body covering, it is necessary to periodically discard the old cuticle to allow further growth. In this case, growth is not a smooth curve (nor even a straight line) but occurs in stages. The body weight gain varies for each molt according to the type of shrimp, age, and environmental conditions (food, water temperature, light) of the habitat. It is usually in the range of 10-50%.

Category 2: Ecdysis occurs in association with regeneration. When the body is injured due to various diseases or when part of the appendages suffer injury or are even lost due to physical factors, ecdysis occurs with regeneration as the chief objective and hence excessive weight gain is precluded. Antennae, thorax, eye stalk, etc. are influenced by the age of the shrimps and environmental conditions. In 2-3 molts they are almost completely normal again.

Category 3: The prespawning molt. Hence it does not occur in juvenile shrimp, immature shrimp, and male parent shrimp; it occurs only in the mature female shrimp. In the case of *M. rosenbergii*, 2-3 days before the prespawning molt the body color changes gradually to light yellowish-brown and the ripe orange ovary is visible through the transparent carapace. The quantity of feeding decreases and the shrimp stops feeding altogether one day before molt. During this period, the new cuticular layer which will become the new exoskeleton after the old one is sloughed is visible, giving the impression of a double layer.

During the process of molt the abdomen undergoes contractions whereby the arthroroidal membrane between the thorax and the abdomen ruptures. First the cephalothorax is molted out followed by molting of the abdomen. Finally the abdomen becomes flexed and the water is expelled by the tail fan and the old exoskeleton is discarded. The duration of molt differs from individual to individual. Longer individuals take 5-10 minutes while shorter ones may complete it in just 40 seconds.

The most important thing about prespawning molt is that it accompanies several morphological changes to make incubation effective. In this context, first a large number of long setae develop at the base of all the pleopods except the 5th pair. These are termed breeding setae and during incubation they become connected with the elastic fiberlike egg stalks and prevent scattering of the eggs. Setae are also found around the genital aperture at the base of the 3rd thoracic legs. Transverse rows of setae likewise develop on the posterior side of the 5th thoracic legs. These setae are called the breeding dress and appear to play some role in assisting the eggs to move into the incubation chamber. On the margin of the I-IV lateral plates of the abdominal exoskeleton also a breeding dress to protect the eggs develops.

Physiological research on the sinus gland and Y-organ is presently underway. Research also includes such aspects as acceleration of molt and spawning, removal of the eye stalk, and administration of ecdysterone and inokosterone.

2.6   COPULATION

The mating behavior of prawns and shrimps differs slightly according to the species. For example, *Penaeus japonicus* displays the type wherein the male and female mate in a horizontal position. In *Palaemon paucidens* and *Penaeus monodon*, mating occurs when the male winds around the female. In another type the male and female are positioned in opposite directions (cephalothorax → tail and tail → cephalothorax) (*Stenopus hispidus*).

In *Penaeus japonicus*, the male shrimp inserts his petasma, the modified part of the 1st abdominal appendage, into the thelycum of the female formed during the prespawning molt. The spermatophores liberated through the genital aperture at the base of the 5th thoracic leg of the male pass through the petasma into the seminal receptacle where they are held together by a sticky substance secreted by it and stored. On the other hand, the chitinous substance derived from the testes is drawn into the accessory genital organ of the female and forms the so-called stopper (Fig. 2.10).

Mating of *Macrobrachium* is almost identical in all species of the genus. Hence the mating behavior of *M. rosenbergii* cultured indoors is described in detail here. In nature, however, the shrimps molt at night but when cultured for a long time in the laboratory, they sometimes molt during the day. Thus a difference in timing was observed. However, no differences in mating behavior were noted.

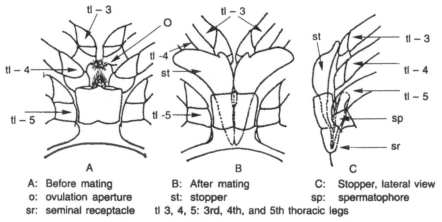

A: Before mating   B: After mating   C: Stopper, lateral view
o: ovulation aperture   st: stopper   sp: spermatophore
sr: seminal receptacle   tl 3, 4, 5: 3rd, 4th, and 5th thoracic legs

Fig. 2.10. Female genital organs of *Penaeus japonicus*.

Usually in the case of prawns and shrimps, the females always molt before mating. This is because preparation for storage of the spermatophores (penaeids) or incubation *(M. nipponense)* is required. Hence this molt is known as the prespawning molt.

In most cases, mating lasts for a few hours from the time the female has undergone prespawning molt; the shrimps have a soft exoskeleton and are not able to creep properly. When male and female are reared separately indoors, unless the molted female is allowed to mate within 15-16 hours of the prespawning molt, the success rate of mating and the percentage of fertilization are both very low. The results of indoor experiments have shown that the success rate is highest between 6-12 hours of mating and almost nil after 20-24 hours.

There are various stages in the mating behavior. First the male identifies the molted female. He spreads out his large right and left 2nd thoracic legs. Using his antennae as feelers he approaches the female and strives to hold her with his legs. While doing so, he expresses disapproval of other males or even females coming too close. He is very possessive of the molted female and even displays a type of protective behavior. This gives the strong impression that after the prespawning molt the female secretes a sex hormone (a pheromone) which reveals her condition to the male.

Next the male mounts and covers the dorsal side of the female. It has sometimes been observed that the female herself turns her body prostrate under the male. In this position the male firmly holds the female with its 3rd, 4th, and 5th thoracic legs, using the distal part of the 3rd thoracic legs to locate the genital aperture in the proximal part of the 3rd thoracic leg of the female.

Generally, after mounting the female from the dorsal side, the male uses its 1st, 3rd, and 4th legs to turn her over and then presses her thorax with the 1st and 2nd abdominal appendages. In this position it slightly flexes the abdominal section and vigorously shakes its body while copulating. This behavior ensures the adherence of the spermatophores to the ventral side of the thorax of the female. The body axis of the male and female *M. rosenbergii* during mating may not always be parallel, in fact it is more often inclined, and hence differs from that observed in *P. japonicus* (Fig. 2.11).

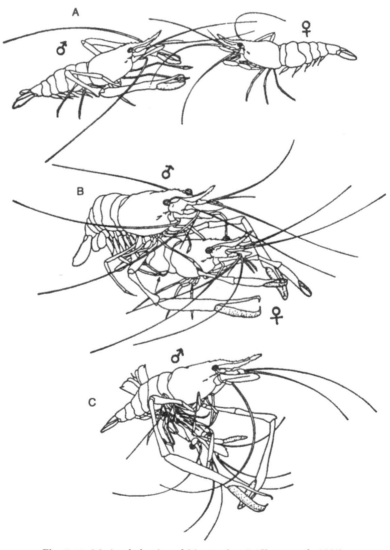

Fig. 2.11. Mating behavior of *M. rosenbergii* (Chow *et al.*, 1981).

The actual duration of mating affecting fertilization is very short, i.e., 2-3 seconds, and the time elapsing from male identification of the procreative female to completion of mating is usually 5-6 minutes, occasionally 10.

As in shrimps and prawns, mating of the American lobster (*Homarus americanus*) also occurs after the prespawning molt of the female. In this species it takes several weeks for the new carapace of the female to completely harden, and sometimes even several months. Mating therefore takes place while the new carapace is still soft and occurs from a few hours to 3-4 days after the prespawning molt. Under natural circumstances, male and female of the same size mate. As in other species, the male in hard exoskeleton, directly introduces sperm into the seminal vesicle in the proximal part of the 4th and 5th thoracic legs of the female. About 30 minutes elapse from the time the male identifies a prespawning molted female to introduction of spermatozoa. Actual mating lasts less than 5 minutes. The mating behavior is almost identical to that in the genus *Macrobrachium*. However, in this case the male and female maintain a parallel and straight position (female infra) during mating. Incubation after coupling begins about 2-3 months later. The sperm stored in the seminal vesicle are retained until the next year's molt.

The mating of *M. nipponense*, *M. formosense* and *M. lar* takes less time than that of *M. rosenbergii*. The shrimps and prawns of *Macrobrachium* and *Penaeus* mate in terms of 1 : 1 (male : female). Contrarily, the *Caridina japonica* female displays mating behavior with several males. In the mating of *M. rosenbergii*, when the male is very small vis-a-vis the female, she often assumes the superior position. Moreover, mating is repeated several times.

2.7 SPERMATOPHORES

In the case of *Macrobrachium*, sperm and gelatinous substance are discharged through the vas deferens, one on either side, and the gelatinous substance binds them into a spermatophore. Using the 1st and 2nd abdominal appendages, the male makes the spermatophore adhere to the thorax of the female between the 2nd and 4th thoracic legs (Fig. 2.12).

The size of the spermatophore of *M. rosenbergii* varies according to the size of the male. In a male in which the carapace is 38.0-68.5 mm, the spermatophore is 6.4-9.8 × 2.4-5.5 mm. Spermatophores adhering to the thoracic plates were removed, cut into transverse sections according to the common method, and stained with Delafield hematoxylin and eosin. When these sections were observed, three parts were seen—one strongly stained with eosin, another weak in basic stain, and a third

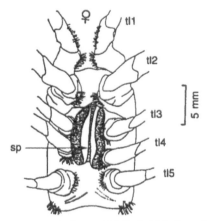

sp: spermatophore   tl 1-5: 1st-5th thoracic legs

Fig. 2.12. Spermatophore adhering to the thorax of female *Macrobrachium rosenbergii* after mating.

where the right and left sperms combined (combining part) (Fig. 2.13). That part adhering to the thorax of the female stained strongly with eosin. Since it has strong adhesiveness soon after discharge during mating, it ensures adherence and fixation of the spermatophore. On the other hand, the weak basic staining substance protects the combining part.

Soon after adherence, the spermatophores are soft, but then gradually swell and in about 10 minutes the adhesive substance solidifies while, contarily, the protective substance softens. The adhesive substance of spermatophores 4-5 hours after mating differs from what it was at the time of adherence. It is weak in eosin staining. The inflated adhesive substance enters into the spaces in the coxae of each thoracic leg and firmly attaches the spermataphores. On the other hand, the protective substance develops a large number of pores compared to its state soon after adherence and the sperms are easily separated.

Spermatophores were collected once during the course of incubation when the eggs were made to adhere to the coxae of the pleopods and again after completion of incubation. When sections of these samples were observed, it was found that there was no change in the adhesive substance and only a small residue of the protective substance was left behind in the samples collected during the course of incubation behavior. Here sperm groups were also less (Fig. 2.12). In the other case, that is, the spermatophores collected upon completion of incubation, there was hardly any protective substance and very few sperm had been left behind.

The structure of the spermatophore is more or less the same in *M. nipponense*, *M. latidactylus*, and *P. paucidens* but the adhesive substance of *M. latidactylus* is highly porous.

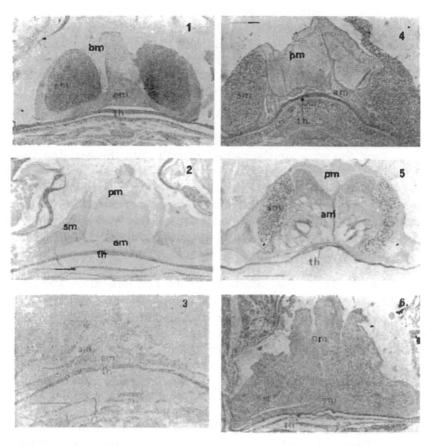

1. Spermatophore adhering to thorax of female *M. rosenbergii* soon after mating (scale length 0.5 mm)
2. Spermatophore adhering to thoracic plate of female *M. rosenbergii* 7 hours after mating (scale length 0.6 mm)
3. Spermatophore adhering to thoracic plate of female *M. rosenbergii* at time of completing spawning (scale length 0.6 mm)
4. Spermatophore adhering to thoracic plate of female *M. nipponense* soon after mating (scale length 0.2 mm)
5. Spermatophore adhering to thoracic plate of female *M. latidactylus* soon after mating (scale length 0.2 mm)
6. Spermatophore adhering to thoracic plate of female *Palaemon paucidens* soon after mating (scale length 0.2 mm).

as: adhesive substance
am: adhering matrix
bm: weak basic staining matrix
em: strong eosin staining matrix
pm: protective matrix
th: thoracic plate
sm: sperm group

Fig. 2.13. Sections of a spermatophore (Chow *et al.*, 1981).

The sperm of shrimps and crabs are highly variable, especially in shape. The sperm of all members of order Decapoda lack a long flagellum at the caudal end and are not motile. The sperm of *M. rosenbergii* is shaped like an alpin. The disk-shaped head has a slight constriction in the center and in cross section appears Y-shaped. The diameter of the head section is 10 microns and the length up to the tip of the tail about 12 microns.

2.8   SPAWNING AND FERTILIZATION

The spawning of prawns and shrimps is broadly classified into 2 types. In one type the eggs are discharged directly into the sea. The sperm received from the male during mating prior to spawning and stored for some time within the body, are also discharged simultaneously. Fertilization occurs in the sea. In the second type the eggs spawned adhere to the bristles on the coxae of the abdominal appendages (swimming appendages) and these bristles grow during the prespawning molt and fertilization occurs here. In this case the eggs are incubated and protected up to hatching.

Penaeus japonicus is a representative of the first type of spawning; unfortunately, there are no reports on its natural spawning habitats. However, when mother prawns with well-developed gonads were cultured, 50-70% of them spawned even in water tanks. Hudinaga (1942) and several other authors have studied the spawning conditions, hatching, metamorphosis of larvae, food, growth, and metabolism under indoor culture.

P. japonicus during spawning discharges eggs while swimming around from midnight to early morning. Each spawn lasts 3-4 minutes. Natural spawning locations of *P. japonicus* are situated in open seas at a considerable distance from the coast. A few reports are available on the number of mature eggs and number of eggs spawned per spawn. The number of eggs increases with an increase in prawn size. For example, a mother prawn of 20-23 cm body length is said to lay 600,000-800,000 eggs.

Reports regarding the fertilization phenomenon of prawns and shrimps are very few. Hudinaga (1942) has detailed this phenomenon in *P. japonicus* (Figs. 2.14 and 2.15). According to him, the unfertilized eggs discharged into the sea form a gelatinous layer containing the ovum. The gelatinous layer is secreted by the mature ovum at the time of spawning (see Fig. 2.7). As a consequence the diameter of the egg decreases to 0.24 mm. The eggs enveloped in the gelatinous membrane remain floating for a short time. When sperm penetrate the gelatinous membrane and come into contact with the egg surface, the fertilization cone is formed. When several sperm come into contact with the egg surface, an equal number of fertilization cones are formed. But the cytoplasmic flow into these cones is not simultaneous. Only the sperm which initially comes into

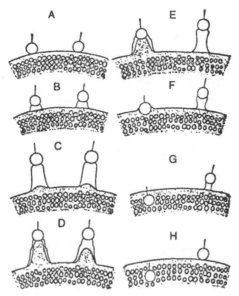

A: Sperm comes into contact with egg surface
B: Shape of fertilization cone
C: Maximum entry of sperm into fertilization cone and expansion of cytoplasm
D: Cytoplasm in contact with sperm head
E: Contraction of fertilization cone
F, G, H: Only one sperm pierces the ovum; the rest remain on its surface

Fig. 2.14. Schematic presentation of fertilization of a shrimp egg (*Penaeus japonicus*) (Hudinaga, 1942).

contact with the cytoplasm enters the ovum and other sperm are left behind on the egg surface. The egg then undergoes maturation division and releases the 1st polar body. This is followed by dilatation of the fertilization membrane. A little later the second maturation division occurs releasing the 2nd polar body. Thereafter the gelatinous layer disappears and is replaced by the egg membrane. The ovum undergoes division. In this stage the ovum gradually sinks (sinking eggs). Species of *Sergestes* lay floating eggs.

American lobster (*Homarus americanus*), European lobster (*Homarus gammarus*), nephrops, rock lobsters, spiny lobsters, and shrimps of Caridae belong to the second type of spawning.

The spawning period of American lobster in America is June to September and even when it was transplanted to Japan this period remained more or less the same. The eggs are discharged through a pair of small genital apertures at the proximal part of the 2nd thoracic (walking) legs. As the eggs pass over the seminal vesicles where the sperm received earlier (a few months before) from the male remain stored, they come

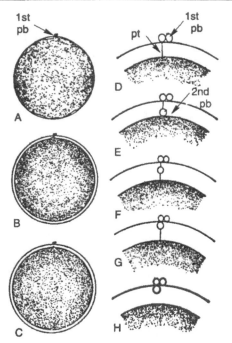

A: Expansion of fertilization membrane
B: Cytoplasm separates from the fertilization membrane and becomes spherical but remains attached by protoplasmic thread
C and D: More or less the same
E: 2nd polar body appears on the protoplasm
F: 2nd polar body approaches 1st polar body
G: 2nd polar body adheres to the fertilization membrane
H: Anchoring thread disappears

Fig. 2.15. Expansion of the fertilization membrane and appearance of 2nd polar body (Hudinaga, 1942).

into contact with the sperm and fertilization occurs. The fertilized eggs adhere to the proximal parts of the abdominal appendages. It takes several hours to complete spawning. The incubation behavior of lobster differs from that of shrimps and prawns. In lobster the 1st pair of thoracic legs (with pincers) is turned backward and the abdomen is flexed. The fertilized eggs are protected from scattering and remain attached in the spaces between the walking legs (Fig. 2.16, A). Prawns and shrimps belonging to this second type of spawning display slight variations in posture during incubation depending on the species.

The egg size compared to that of *Penaeus japonicus* is slightly larger, about 1.6 mm in diameter. As embryonic development progresses the egg elongates and its color changes from dark brownish-green to a lighter hue. The number of incubating eggs per lobster varies according to the

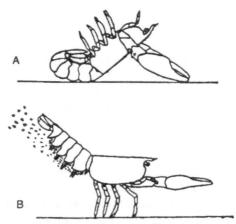

A: Posture during egg incubation   B: Posture during release of hatched larvae

Fig. 2.16. Incubation and hatching in American lobster.

size of the mother lobster. It is putatively about 5,000 eggs when the body length is 20 cm, 20,000 eggs at body length 30 cm, 40,000 eggs at body length 36 cm, and 63,000 eggs at body length 43 cm. The number of eggs is generally reported as 3,000-75,000 or 5,000-115,000.

The duration from spawning to hatching is greatly influenced by water temperature. The incubation period is long in those species living in cold regions and deep seas. In the American lobster, which lives in cold seas where the water temperature drops to –1.8°C (moving southward the water temperature in summer rises to 24°C) and is widely distributed from Labrador to North Carolina (USA), the incubation period reportedly ranges from 6 months to even one year. On the other hand, hatching of eggs spawned in summer in Japanese waters where the temperature is fairly constant at 22°C has been reported as starting from November. As a matter of fact, egg hatching commences mid-June, peaks in July, and continues through November. The duration of hatching from beginning to end varies according to water temperature. At a temperature of 12-14°C, there is a variation of about 2 weeks. The behavior of mother lobsters during incubation is more or less the same as that of *Macrobrachium* species. The body is held high with the thoracic legs and the cephalothorax held parallel to the sea floor. The tail fan is raised in such a way that the abdomen acquires an angle of 20-45°. In this position the abdominal appendages are "pedaled" rapidly which causes the eggshells to rupture and the larvae are discharged. The pedaling of the abdominal appendages causes a water current and the larvae are scattered backward (Fig. 2.15, B).

Factors inducing hatching have yet to be worked out. However, continuous observation during indoor rearing suggests that hatching begins from around sunset and around dawn and there appears to be some relationship between initiation of hatching and light. Strong aeration also appears to initiate hatching.

The following text describes the spawning and fertilization of *M. rosenbergii*. This species is a large freshwater prawn considered to be the world's finest grade. The size often exceeds 30 cm. It is widely distributed in Southeast Asian regions including Vietnam, Cambodia, Thailand, Burma, Pakistan, Malaysia, Indonesia, and the Philippines. Efforts are underway in various regions to culture it.

After completing the prespawning molt, the female mates and receives the spermatophores. Within a few hours spawning begins. Just before spawning she briskly cleans the breeding dress on the thoracic part of the body and along the margins of the lateral plates of the abdomen, as well as the breeding setae in the coxae of the 1st to 4th abdominal appendages. Cleansing is done with the 1st pair of thoracic legs. Sometimes she folds the abdominal appendages forward and continues to swim.

Once spawning has started, the body is held slightly raised by means of the thoracic legs and the caudal part is bent downward. All the abdominal appendages are folded forward. Of these, the 1st pair are folded in such a manner that they cover the genital (spawning) apertures located in the proximal parts of the 3rd thoracic legs. The discharged eggs pass through the spermatophores adhering to the 1st abdominal legs and thorax and reach the incubation chamber. The incubation chamber or the brood pouch is divided into compartments, No. 1, No. 2—No. 5 but there are no breeding setae in the proximal parts of the 5th abdominal appendages nor is there any breeding dress along the margins of the lateral plates of the abdomen. Therefore incubation does not occur in the 5th compartment, which serves merely as a vestibular chamber. The eggs migrating continuously become attached to the brood pouch in a sequential manner from compartment No. 4 to No. 1. During this process the female shakes her body right and left every few seconds to make the eggs spread in a uniform manner.

The egg masses inside the brood pouch are covered by an adhesive substance. These masses were fixed and tissue sections prepared and examined (Fig. 2.17). Within the adhesive substance, there were many sperm mixed with eggs. Observations also showed various states of fertilization such as sperm in contact with the surface of the egg membrane, sperm penetrating the egg membrane, and sperm which had entered the ova (Fig. 2.18). In view of these observations, it appears that in the case of *Macrobrachium rosenbergii* fertilization occurs during the course of the eggs emerging from the genital apertures proceeding to the incubation chamber, where some spermatophores adhere to the eggs, resulting in the union of ova and sperm.

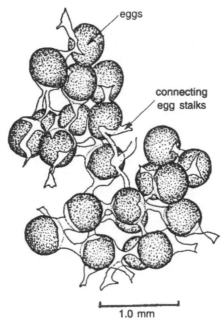

Fig. 2.17. Fertilized eggs in the incubation chamber (Kwon, 1982). Adhesive fluid adheres to the eggs, coagulates and turns into threadlike stalks of the eggs, which are then interconnected.

When *M. rosenbergii* cultured indoors was continuously examined from prespawning molt to mating followed by spawning, it was found that unless mating occurs within 24 hours after the prespawning molt, fertilization will not result. Most of the cultured mother prawns are found to undergo prespawning molt late in the night. When these females are allowed to mate from the following morning to noon, most of them spawn within about 12 hours. Even when not mated, they spawn within about 24 hours. When the spawning of such unfertilized eggs is delayed by 2-3 days, they undergo prespawning molt once again within one week. In these prawns the intervals between successive prespawning molts are shorter than in the normal case.

The number of incubating eggs in *M. rosenbergii* is smaller than that of *kuruma* shrimp or spiny lobsters. It is 40,000-60,000 in a still growing prawn (less than one year in age, body length 15-16 cm and weight 70-80 g) and about 100,000 eggs in a 1.5-2-year-old (body length 18 cm and weight 100 g). A young female 6-7 months old lays 4,000-10,000 eggs. Even when these prawns are grown in small water tanks, if the water temperature is maintained at 27-28°C they spawn 4-7 times (maximum recorded to date, 8 times) in a year. Thus they are termed high-proliferation prawns.

In the case of *M. rosenbergii*, the prespawning molt is sometimes repeated 8 times at intervals of 16.9 ± 1.6 days at a water temperature of 25.9 ± 2.1°C. In *M. formosense*, there are cases in which the prespawning molt is repeated 7 times at intervals of 18.6 ± 5.4 days and water temperature 28.3 ± 1.3°C. Again, in *M. rosenbergii*, if hatching of incubating eggs occurs in the early and middle periods of spawning, the next prespawning molts occurs the same day.

All members of *Macrobrachium* move to fresh waters during the juvenile stage. However, while rearing the incubating mother prawns it is suggested that a little sea water be added as this increases the hatching rate and is also effective in disease prevention. A chlorine concentration of 2-5‰ is best suited in the rearing of *M. rosenbergii*.

2.9 EMBRYONIC DEVELOPMENT AND HATCHING OF *MACROBRACHIUM* PRAWNS

At the time of spawning, the eggs of *M. rosenbergii* are elliptical in shape and 0.58 × 0.48 mm in size. They are opaque and pale orange. At a water temperature of 28.0 ± 0.4°C, within 5-6 minutes of fertilization the 1st polar body is discharged and the fertilization membrane expands. The 2nd polar body is then discharged. Up to this stage, egg development is more or less similar to that of *kuruma* shrimp studied by Hudinaga (1942). However, the eggs of *M. rosenbergii* next undergo nuclear division twice and attain a 4-celled stage. Cleavage then becomes visible. Until then the cytoplasmic division cannot be clearly seen under the microscope. While the eggs of *Macrobrachium* prawns are of the superficial-cleavage type, those of *kuruma* shrimp are of the equal-cleavage type.

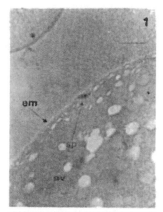

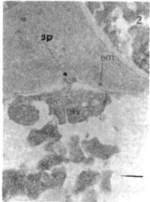

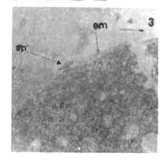

1. Sperm (sp) seen on egg membrane (em)
2. Sperm (sp) seen on egg membrane (em)
3. Expansion of fertilization cone

Fig. 2.18. Fertilization of *M. rosenbergii* eggs (Chow *et al.*, 1981).

Beyond the 4-celled stage, nuclear divisions can be readily observed along with cytoplasmic division under the microscope. At a water temperature of 27-28°C, they attain the morula stage in about 20 hours after fertilization. Thereafter development passes through the blastular stage, the gastrular stage, and closing of the blastopore begins about 54 hours later. Then begins the determination of various rudiments and subsequent differentiation. After 80 hours, the embryo appears. After 170 hours heartbeats are discernible and after 230 hours the compound eyes develop. From this stage onward the various organs of the body gradually form. After 320 hours the embryo assumes the shape of a zoea. From this time the egg axis increases and after 430 hours has attained maximum length. The embryo is now 0.85 × 0.54 mm and ready to hatch (Fig. 2.19).

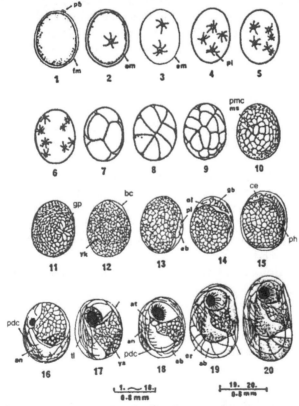

pb:1st and 2nd polar bodies; fm: fertilization membrane; em: egg membrane; pi: protoplasm islands; pmc: primordial mesodermal and endodermal cells; bp: blastopore; yk: yolk; bc: blastocyst; rb: rudimentary brain; ol: optical lobe; eb: embryo; ce: compound eye; ph: primordial heart; pdc: primordial digestive canal; an:1st antenna; at: 2nd antenna; tf: tail fan; tl: thoracic legs; cr: carapace; ab: abdominal segments

Fig. 2.19. Embryonic development of *Macrobrachium rosenbergii* (Kwon, 1982).

When, subsequent to mating, mother prawns with incubating eggs are reared in water tanks with a 3.12-4.28‰ chlorine concentration and a temperature of 26.0°C, the larvae hatch after 25 days. When the water temperature is 28.0-28.5°C, they hatch in 20 days and at 32.0-32.2°C in 17 days. From the results of experiments, replicated 20 times, the relationship between the required duration from the time of fertilization (Y) and water temperature (X) can be expressed by the following equation:

$$\log Y = \log 6026.6 - 1.69945 \log X.$$

There are still many aspects requiring investigation. However, from the hatching rate, variations in timing of hatching, and the survival percentage of zoea, it appears that the suitable conditions for rearing pregnant mother prawns are water temperature maintained around 28°C and the environment so controlled that hatching is ensured in about 20 days. The pattern from embryonic development to hatching varies slightly in *Macrobrachium* depending on the species, but is nevertheless quite similar.

2.10   LARVAE OF SPRING LOBSTER AND *KURUMA* SHRIMP

Depending on the species of shrimps and prawns, hatching larvae differ widely in general morphology and structure and function of various parts. These differences are not fundamental, but attributable to the difference in developmental stages at the time of hatching. The larvae can thus be classified into four types according to the stage at which they hatch: early stage, slightly advanced stage, or fully grown stage. The four types are:

1) Larvae in early stage called nauplius with simple form hatch.
2) Larvae pass the nauplius stage inside the egg and hatch in the next stage termed zoea.
3) Both nauplius and zoea stages are passed inside the egg and the larvae hatch as mysis.
4) All larval stages are passed inside the egg and fully grown forms resembling the adult hatch.

The species of prawn whose eggs are large and contain a large quantity of yolk hatch at an advanced stage of growth and the duration of embryonic growth from the time of fertilization is long.

Larvae of spring lobsters: *Palinurus* types hatch as a phyllosoma—a larval form with unique features. This corresponds to the early stage of mysis. After repeated molts and growth, it transforms into a form called the puerulus whose morphological features and appendages resemble those of the adult. The puerulus is also popularly known as glass shrimp. After further molts it becomes a juvenile prawn. So far it has not been possible to rear it up to the juvenile stage and hence many details are still not known.

South African genus *Jasus* resembles a phyllosoma at the time of hatching. It attains this stage after passing through 2 stages, namely, the prenaupliosoma and the naupliosoma.

Based on the research of many experts, duration from hatching to the juvenile stage and number of stages during this period are given in Tables 2.5 and 2.6. It can be seen from these Tables that there are species which take less than a month to attain the juvenile stage and others which may take one or nearly two years. Compared to other prawn species, the larval period of *Palinurus* is usually longer. Due to difficulties relating to artificial diet, details of the early period of its history have yet to be clearly understood.

Table 2.5: Larval period of spiny lobsters

Type	Larval period	References
Family Palinuridae		
Panulirus argus	6 months	Lewis, 1951; Sims and Ingle, 1996
P. interruptus	7.75 months	Johnson, 1971
P. penicillatus	probably 7-8 months	Johnson, 1971
P. cygnus	9-11 months	Chittleborough and Thomas, 1969
P. homarus	4-6 months	Berry, 1974
P. japonicus	perhaps 8-10 months	Inoue, 1981
Palinurellus gundlachi	10 months	Sims, 1966
Jasus lalandii	9-10 months	Lazarus, 1967
J. edwardsii	12-22 months	Lesser, 1978
Family Scyllaridae		
Scyllarus americanus	32-40 days	Robertson, 1968
S. depressus	25 days	Robertson, 1968, 1971
S. chacei	about 6 weeks	Robertson, 1968
S. arctus	probably 3-4 months	Robertson, 1968
S. bicuspidatus	probably 3-4 months	Robertson, 1968
S. planorbis	54 days	Robertson, 1979
Scyllaridis aequinoctialis	perhaps 8 or 9 months	Robertson, 1969
S. nodifer	9 months	Sims, 1965; Robertson, 1969

As shown in Fig. 2.20, the phyllosoma of spiny lobsters has a fragile body and appears to be a weak swimmer. Perhaps it is influenced by the water current and is widely scattered. Some larvae are found floating in their habitats while others settle to the bottom. By and large, little is known about the distribution of larvae in nature, their food, or pattern of living.

At the fishery experimental stations of Kanagawa, Chiba, and Shizuoka prefectures, intensive research has been carried out on various aspects of artificial seedling production of spiny lobsters. These studies include

Table 2.6: Number of larval stages of spiny lobsters

Type	Number of larval stages	References
Family Palinuridae		
Panulirus argus	11	Lewis, 1951
P. interruptus	11	Johnson, 1956
P. penicillatus	10	Prasad and Tampi, 1959
P. cygnus	9	Brain et al., 1979
P. homarus	9	Berry, 1974
P. japonicus	11	Inoue, 1978, 1981
Palinurellus gundlachi	12	Sims, 1966
Jasus lalandii	13	Lazarus, 1967
J. edwardsii	11	Lesser, 1978
Family Scyllaridae		
Scyllarus americanus	6-7	Robertson, 1968
S. depressus	9-10	Robertson, 1971
S. chacei	6-7	Robertson, 1968
S. arctus	9	Robertson, 1923; Santucci, 1925 Kurian, 1956
S. bicuspidatus	10	Saisho, 1966, Robertson, 1968
S. planorbis	8	Robertson, 1979
Scyllaridis aequinoctialis	11	Robertson, 1969
S. nodifer	12	Sims, 1965; Robertson, 1969

a report on the results of rearing phyllosoma up to the 11th stage by Inoue (1981), who studied rearing of larvae for several years. According to this report, phyllosoma soon after hatching are 1.53-1.54 mm in length. The report also mentions that the peak period of hatching under natural conditions is earlier in sea waters of the southern region and can be divided into two phases—one early June, the other mid-July to early August. There are mother prawns which spawn twice during the spawning period but the offspring of the second spawn are weak and mortality is high. It is also mentioned that at low salinity the intervals between molts are very long, growth is also poor, and more than 32‰ salinity is required for survival. Inoue further notes that the ideal water temperature for rearing is 26-27°C and nauplius of *Artemia*, larvae of various other crustaceans, fingerlings, larvae of annelids, sea urchin eggs, flesh and liver of fishes, and molasses can be used as feed. However, arrowworms and fingerlings are said to be the best feed.

Larvae of *Kuruma* shrimp: From the pattern of hatching *Kuruma* shrimp belong to type 1 and all of them hatch as nauplius (Fig. 2.21). The body of the nauplius is unsegmented and has 3 pairs of appendages, of which the first pair is uniramous and the other two biramous. The three pairs correspond to the 1st antennae, 2nd antennae, and mandibles of the adults. At the nauplius stage the larvae are active free swimmers but lack functional organs otherwise.

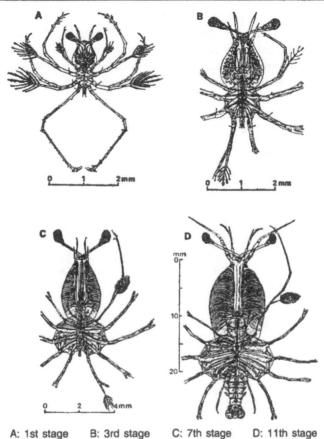

A: 1st stage   B: 3rd stage   C: 7th stage   D: 11th stage

Fig. 2.20. Phyllosoma of spiny lobsters (Inoue, 1981; and others).

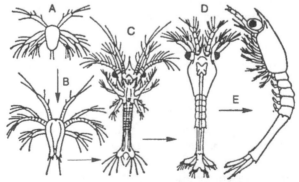

A: Prenauplius   B: Nauplius   C: Zoea   D: Premysis   E: Mysis

Fig. 2.21. Larvae of *Kuruma* shrimp [Illustrated Fauna of Japan (Nihondobutsuzukan) 1965].

The nauplius of *Kuruma* shrimp undergoes 6 stages. From the first stage when it is just 0.3 mm in size it undergoes repeated molts and metamorphoses into the next larval stage, i.e., zoea. As mentioned in Table 2.7, even among members of the genus *Penaeus* there are variations in number of various growth phases (nauplius, zoea, and mysis) depending on the species.

Table 2.7: Number of stages in each growth phase during the larval period of *Kuruma* shrimp

Species	Growth phase			References
	Nauplius	Zoea	Mysis	
Penaeus japonicus	6	3	3	Hudinaga (1942)
P. chinensis	6	3	3	Oka (1967)
P. trisulcatus	8	3	4	Heldt (1938)
P. setiferus	5	3	2	Pearson (1939)
P. setiferus	4-5	3	2	Johnson and Fielding (1956)
P. duorarum	5	3	3	Dobkin (1961)
Metapenaeus monoceros	6	3	3	Hudinaga (1942)
M. burkenroadi	6	3	3	Hudinaga (1942)
M. dobsoni	3	3	3	Menon (1951)
Trachypenaeus constrictus	5	3	2	Pearson (1939)
Xiphopenaeus kroyeri	5	—	—	Renfro and Cook (1963)
Sicyonia brevirostris	5	3	4	Cook and Murphy (1965)
S. wheeleri	3	3	3	Gurney (1943)
S. stimpsoni	5	3	2	Pearson (1939)
S. carinata	8	3	4	Heldt (1938)

During the nauplius larval stage of *Kuruma* shrimp the larvae grow with the yolk as the nutritive source and never feed from an external source. The larva entering the zoea stage undergoes severe morphological transformation and in the second stage of zoea, maxillipeds develop and active feeding commences. The compound eyes fused with the carapace now separate and eye stalks develop. In the third and final stage of zoea, uropods become distinct. The morphological and functional modifications occurring during the zoea stage are more or less similar to those of *Macrobrachium* and *Numa* shrimps.

In the next or mysis stage the thoracic appendages (walking legs) are well developed, in particular the exopodites, as swimming organs. All 5 pairs of thoracic appendages are biramous in the mysis stage and resemble those of Mysidacea. Hence the larvae are called mysis. After passing the mysis stage they are generally termed postlarvae and the morphological features resemble the adult form. Their pattern of living also changes, i.e., from free swimmers they shift to benthic dwellers. In shape as well as in mode of living they still differ from the adult form. In particular, the 6th pair of abdominal appendages are far longer than those of the adult. It is only after 10 or even 20 molts that the postlarva becomes an adult (Table 2.8).

Table 2.8: Growth stages and size of *Kuruma* shrimp larvae (Hudinaga, 1942)

Growth stages		Maximum (mm)	Minimum (mm)	Average (mm)
Nauplius	1st stage	0.34	0.30	0.32
"	2nd stage	0.35	0.33	0.34
"	3rd stage	0.38	0.35	0.37
"	4th stage	0.42	0.38	0.39
"	5th stage	0.45	0.42	0.44
"	6th stage	0.51	0.48	0.50
Zoea	1st stage (pre)	0.93	0.87	0.92
"	(post)	1.32	1.20	1.30
"	2nd stage (pre)	1.65	1.33	1.58
"	(post)	2.13	1.35	1.97
"	3rd stage (pre)	2.32	2.14	2.24
"	(post)	2.59	2.33	2.50
Mysis	1st stage	3.10	2.67	2.83
"	2nd stage	3.64	2.99	3.34
"	3rd stage	4.52	3.79	4.34
Postlarva		5.00	4.79	4.90

The names for the larval stages based on morphological differences and number of stages based on number of molts vary slightly according to the researcher. For example, Gurney (1926, 1942) considered the swimming function and called the stage following the nauplius protozoea. On the other hand, Williamson (1969) suggested that the larvae of prawns and crabs pass from nauplius to zoea and then to megalopa. Again, according to Shiino (1964) there are 2 stages in nauplius, 5 stages in metanauplius, 1 stage in protozoea, 1 stage in zoea, and 3 stages in mysis; after these stages the larva becomes a decapod. Chace (1960) has described the course of larval transformation as follows:

Family Penaeidae: Nauplius → Protozoea → Mysis (zoea)
                      → Mastigops (postlarva)
Family Sergestidae: Nauplius → Protozoea → Acanthosoma (zoea)
                      → Mastigops (postlarva)

Regarding *Penaeus monodon*, Motoh (1979) has given the following description: as in *Kuruma* shrimp, the nauplius has 6 stages, zoea 3 stages and mysis 3 stages. Thereafter they enter the postlarval stage (5.50-6.13 mm body length). In the early period they molt at the rate of one molt per day and after 20 molts become a fully formed adult.

There are many reports on growth stages for several species and some idea of this aspect can be obtained from Table 2.7.

2.11 LARVAE OF *MACROBRACHIUM*

The hatching pattern of *Macrobrachium* falls in the second type described earlier. In other words, the nauplius stage is passed inside the egg and hatching occurs when growth has attained the stage of zoea. In this case also the number of stages during the period of zoea differs from species to species.

According to Holthuis (1950, 1952), there are about 130 species of *Macrobrachium* in the world. Of these, only about 20 species have been studied for early development, early part of life history, or complete life history. Besides *M. rosenbergii* this list includes the following:

Macrobrachium carcinus (Lewis, 1961)

Macrobrachium carcinus (Lewis and Ward, 1965)

Macrobrachium carcinus (Choudhury, 1971)

M. (Palaemon) lamarrei (Rajyalakshmi, 1961)

M. nipponense (Kwon and Uno, 1969)

M. formosense (Shokita, 1970)

M. acanthurus (Choudhury, 1970)

M. acanthurus (Dobkin, 1971)

M. australiense (Fielder, 1970)

M. intermedium (Williamson, 1972)

M. niloticum (Williamson, 1972)

M. shokitai (Shokita, 1973)

M. japonicum (Kwon, 1974)

M. hendersodayanum (Jalihal and Sankolli, 1975)

M. olfersii (Dugger and Dobkin, 1975)

M. americanum (Monaco, 1975)

M. equidens (Ngoc-Ho, 1976)

M. novaehollandiae (Greenwood et al., 1976)

M. asperulum (Shokita, 1977)

M. lar (Atkinson, 1977)

M. amazonicum (Guest, 1979)

Complete details from the first-stage zoea to postlarva have been studied in only half of the species listed above.

M. rosenbergii is one of the best covered species, having been studied by S.W. Ling, Fujimura, and several other researchers. *M. rosenbergii* passes through 11 stages of zoea to transform into a postlarva (juvenile). It attains the juvenile stage in 31-33 days (Table 2.9) when the water temperature is 27.5±0.5° C and the salinity 5-7‰, and in 22-28 days when reared in open tanks under shade in summer (24-33°C).

Table 2.9: Growth of *Macrobrachium* larvae reared at 27.5 ± 0.5°C water temperature and 6.58-6.81‰ salinity

Stages	Duration (days) up to each stage		Body length (mm)		No. of specimens
	Average	Range	Average	Standard deviation	
Zoea 1	1	1–2	1.92	0.02	15
" 2	3	2–4	1.99	0.06	12
" 3	5	4–7	2.14	0.05	12
" 4	8	6–10	2.50	0.08	14
" 5	10	9–13	2.84	0.07	14
" 6	14	12–18	3.75	0.37	19
" 7	17	15–20	4.06	0.15	13
" 8	20	18–23	4.68	0.20	12
" 9	24	21–29	6.07	0.29	17
" 10	28	25–34	7.05	0.52	23
" 11	31	28–37	7.73	0.81	21
Postlarva	34	31–43	7.69	0.65	50

The first-stage zoea is about 1.9 mm in body length, 0.5 mm in carapace length, and the head is bent downward. It actively swims around in an inverted position. In this stage the eyes are fused with the carapace and lack stalks and the telson is not differentiated from the uropod.

The eye stalks develop in the 2nd stage and the rudimentary tail fan appears, as do the appendages. The major changes include the complete consumption of yolk and an indication of feeding behavior. In the 3rd stage all the larvae begin to feed.

During the zoea stage the larvae feed mostly on nauplius larvae of *Artemia*. In the early zoea stages even some rotifera such as *Brachionus* serve as food. When the zoea has grown to the 4th-5th stage, it feeds on fish meat and molluscan flesh. Since the larvae in the zoea stage are free swimming and do not migrate to the bottom, food items which rapidly settle down are rarely consumed.

When it became possible to produce seedlings of *M. rosenbergii*, Ling (1969) reported the following: Yolk particles prepared by sieving boiled yolks through 25 mesh cm^{-1} can be used as feed for early larvae 2-4 days after hatching (when zoea start feeding); the same sieved through 20 mesh cm^{-1} is suitable for 5-10-day-old larvae; the same sieved through 12 mesh cm^{-1} for 11-20-day-old larvae; and that passed through a 7 mesh cm^{-1} for larvae more than 20 days old. Whole eggs of various fishes also proved effective feed.

On the other hand, Morimi (1973) obtained a rich harvest of seedlings by feeding early-stage larvae with rotifers such as *Brachionus* (he never gave nauplius of *Artemia*) and later with clam meat.

In the seedling production of fisheries, the size of the feed provided is correlated with size of the fingerlings, a very important point indeed. But in the case of prawn larvae, since the food is grabbed and crushed by the maxillipeds, the feed can be larger in size. These crustaceans are polyphagous and strongly inclined to feed on a wide variety of food species.

Brightness also exerts a strong influence on acceptance of feed. In *M. rosenbergii* the rate of feeding by various stages of zoea increased rapidly concomitant with an increase in brightness in the range of 0 lux to 4,000 lux. In zoea stages 8-10 compared to zoea of stages 2-4, the ratio of increase in feeding due to brightness was higher. Among zoea of the same stage, the maximum increase in feeding rate was found between 0-500 lux. At a brightness of 7,000 lux, feeding reduced to half that at 4,000 lux. When the larvae attained the juvenile stage, the tendency toward increased feeding in relation to brightness was negligible. After metamorphosis, many physiological changes seem to occur.

In *Macrobrachium*, the feeding rate was maximum at 4,000 lux and minimum at 0 lux. This trend is evident in *M. rosenbergii*. Even at 0 lux *Macrobrachium* feed but compared to *M. rosenbergii*, the ratio of increase in feeding concomitant with increase in brightness from 0-4,000 is lower.

Uno *et al.* (1967) studied the relationship between feed density and feeding rate in larvae of *M. nipponense*. According to these authors, the relationship can be applied to the theoretical equation of Ivrev, that is, $-r = R (1 - e^{-kp})$. The feeding rate r of 9th-stage zoea at 28°C water temperature can be expressed as

$$-r = 22.63 (1 - e^{-0.3440p})$$

where p is feed density. It was reported that the minimum feed density at which the 9th-stage zoea can be initiated to feed is 1,200 *Artemia* larvae/ 300 ml culture medium.

M. rosenbergii and various other species of *Macrobrachium*, including the Mexican species *M. americanum* and *M. carcinus*, migrate to fresh water habitats after the postlarval stage. They are popularly known as freshwater prawns. When the larvae are reared indoors, the rearing water requires a little salinity. If pure fresh water or pure sea water (with more than 19‰ salinity) is used, molting is delayed, growth is poor, and mortality is high. Most of these factors have not yet been clearly understood. However, it is clear that there is a significant difference in the desirable chlorine concentration suited for early-stage zoea and late-stage larvae.

For *M. rosenbergii*, the culture medium prepared by diluting the sea water to 30-40% is suitable (chlorine content 5.7-7.7‰). This species has very low resistance to fresh water. The oxygen consumption also declines if the chlorine concentration falls below 1.9‰ or increases above 19.0‰.

For *M. nipponense* the chlorine concentration may be in the range of 1.9 to 7.7‰. It is strongly resistant to fresh water. On the contrary, it has very poor resistance to 100% sea water (19.1‰ chlorine).

For *M. formosense* the chlorine concentration is best suited in the range of 5.7-7.7‰. This species is strongly resistant to both fresh and sea water.

The desirable chlorine concentration for *M. japonicum* is 5.7-9.6‰; further, this species has strong resistance to fresh water and can withstand a wide range of salinity.

The desirable salinity range for *M. americanum* and *M. carcinus* is the same as that of *M. japonicum* but the suitable range of fresh water and sea water is relatively narrow.

M. rosenbergii undergoes repeated molts at intervals of about 3 days after hatching. After molting 11 times, the body length attains 8 mm and it becomes a juvenile prawn. The speed of growth of this period is gradually influenced by the temperature of the culture medium, chlorine concentration, quality and quantity of feed, and quality of rearing water. The results of experiments in which the prawns were reared at 27.6-28.8°C water temperature, 6.71-6.95‰ chlorine concentration, 8.0-8.2 pH, fed with nauplius of *Artemia* at a density of 5-7 larvae cc^{-1} could be expressed by the equation

$$Y = 0.8533 \ e^{0.3706X}$$

where X is the number of stages and Y the mean growth of each stage (Fig. 2.22). In the same way, the carapace growth can be expressed as $Y = 0.3718 \ e^{0.1478X}$ and the abdominal growth as $Y = 0.1636 \ e^{0.1540X}$. In other words, it matches well with the growth formula of Brook, i.e., $Y = Be^{aX}$.

As for the morphological changes at each stage from first stage zoea to the juvenile stage, detailed reports have been given by Uno and Kwon (1969) and Ling (1962, 1969) (Fig. 2.23).

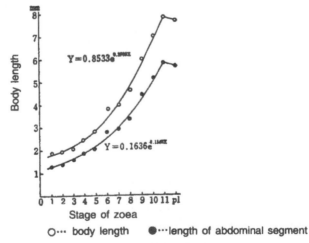

Fig. 2.22. Growth of *M. rosenbergii* larvae reared at 28.5 ± 0.5°C water temperature and 6.58-6.81‰ chlorine concentration.

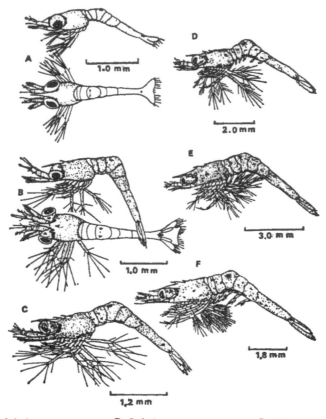

A: 1st-stage zoea   B: 3rd-stage zoea   C: 5th-stage zoea
D: 7th-stage zoea   E: 9th-stage zoea   F: 11th-stage zoea

Fig. 2.23. Growth and morphological changes of *M. rosenbergii* larvae (Kwon, 1982).

REFERENCES

Atkinson JM. 1977. Larval development of a freshwater prawn, *Macrobrachium lar* (Decapoda, Palaemonidae), reared in the laboratory. Crustaceana, 33 (2).

Cobb JS, Phillips BF. 1980. The Biology and Management of Lobsters, vols. 1 and 2. Academic Press, London-NY.

Fujinaga M, Kotaka J. 1964. Metamorphosis and feed of *Penaeus* larvae. Plankton Res. Bull. Japan, 13.

Fushimi, K. 1976. Ecology of spiny lobster of southern Izu region. Suisan Doboku, 12 (2).

Guest, WC. 1979. Laboratory life history of the Palaemonid shrimp *Macrobrachium amazonicum* (Heller). Crustaceana, 37 (2).

Hanson JA, Goodwin HL. 1977. Shrimp and prawn farming in the western hemisphere. Dowden Hutchinson and Ross, NY.

Harada, E. 1965. Spiny lobster and its geographic distribution. Nankiseibutsu, 7 (2).

Hudinaga M. 1942. Reproduction, development and rearing of *Penaeus japonicus* Bate. Jap. J. Zool., 10 (2).

Ichiku S. et al. 1976. Ecology of puerulus larvae and early postlarvae of spiny lobster. Marine Eng. (Suisan do bo ku) 12(2).

Ikeda A. 1962. Catching of *Penaeus chinensis* of the Yellow Sea. Seisukenpo, 27.

Ikematsu T. 1963. Life history of shrimps and mysids of the Ariake Sea. Seikaiku Suikenpo, 30.

Ikematsu T. 1967. Study of *Penaeus chinensis* culture, I. Suisan Zoshoku, 15 (2).

Ikematsu T., Kimura S., Yamashita K. 1967. Study of *P. chinensis* culture, III. Suisan Zoshoku, 15 (2).

Inoue S. 1964. Feeding rate of spiny lobster during its culture. J. Fisheries Assoc. Japan (Nisuishi), 30 (5).

Inoue S. 1965. Feed of early stage phyllosoma of spiny lobster. J. Fisheries Assoc. Japan (Nisuishi) 31 (11).

Inoue S. 1981. Fundamental study of the rearing of phyllosoma larvae of spiny lobster. Kanagawa Suisanshironshu, 1.

Inoue A. 1949. Ecology of *Leander serrifer* Stimpson. J. Fisheries Assoc. Japan (Nisuishi), 15 (7).

Iwamoto, K., Mizue I. 1968. Numa shrimp culture, II. Suisan Zoshoku, 15 (4).

Kinoshita T. 1934. Puerulus of spiny lobster and changes thereafter. J. Zoonal (Dobutruzatsa), 46: 551.

Kitani H., Alvarado JN. 1982. The larval development of the Pacific brown shrimp *Penaeus californiensis* Holmes reared in the laboratory. Bull. Jap. Soc. Sci. Fish., 48 (3).

Kobayashi S. 1955. Spawning and growth of freshwater giant prawn of Lake Biwa. Bull. Fisheries Exper. Station Shiga, 8.

Kubo I. 1948. Propagation of freshwater shrimps. Bull. Jap. Soc. Sci. Fish. (Suiken Kaiho), 1.

Kubo I. 1949. Propagation of freshwater shrimps. Bull. Jap Soc. Sci. Fish. (Suiken Kaiho), 2.

Kubo I. 1949. Oecological studies on the Japanese freshwater shrimps, *Palaemon nipponensis* 1. Seasonal migration and monthly size-composition with special reference to growth and age. Bull. Jap. Soc. Sci. Fish., 15 (2).

Kubo, I. 1950. Ecological study of freshwater giant prawns. J. Fisheries Assoc. Japan (Nisuishi), 15 (10).

Kubo, I. 1950. Study of propagation of freshwater shrimps. J. Fisheries Res. (Suikenkaiho), 3.

Kurata S. 1971. Biology of *Penaeus japonicus*. Kosei Pub., Tokyo.

Lewis JB. 1951. The phyllosoma larvae of the spiny lobster, *Panulirus argus*. Bull. Mar. Sci., Gulf and Caribbean, 1.

Lynn John, W., Clark Wallis, H. Jr. 1983. A morphological examination of sperm-egg interaction in the freshwater prawn, *Macrobrachium rosenbergii*. Biol. Bull., 164 (3).

Lynn John, W., Clark Wallis, H. Jr. 1983. The fine structure of the mature sperm of the freshwater prawn, *Macrobrachium rosenbergii*. Biol. Bull., 164 (3).

Matsui S. and Waiychi K. 1937. Ecological studies of *Leander paucidens* de Haan of Lake Towadaj. Rikusui Gakkaiho, 7: 1.

Mizue I., Iwamoto K. 1960. Shrimp culture. Suisan Zoshoku, 8 (12).

Mori M., Nakamura S. 1973. Study of seedling production of *M. rosenbergi*. Suisan Zoshoku, 21 (3).

Morokita S. 1966. Ecology of *M. nipponense* and larval metamorphosis. J. Biol. Assoc. Okinawa (Okinawa Seibutsu Gakkaishi), 3 (5).

Morokita S. 1970. Study of *Macrobrachium formosense*, I. Larval changes in indoor culture tanks. J. Biol. Assoc. Okinawa (Okinawa Seibutsu Gakkaishi), 6 (8).

Motoh H. 1981. Studies on the fisheries biology of the giant tiger prawn, *Penaeus monodon* in the Philippines.

Murano M. 1967. Preliminary notes on the ecological study of the phyllosoma larvae of the Japanese spiny lobster. Bull. Planktology Jap. Commemorative issue for Dr. Y. Matsue.

Nagai S. 1956. Mating and spawning of spiny lobster. Suisan Zoshoku, 4 (2).

Ohshima K. 1936. Feeding during early stage phyllosoma of spiny lobster. Bull. Fish. Assoc. (Suisan Gakkaiho), 7 (1).

Ohshima K. 1946. Certain aspects of ecology of spiny lobster. Bull. Fish. Assoc. (Suisan Gakkaiho), 8 (3-4).

Ohshima K., Yasuda J. 1942. Ecology of *Penaeopsis affinis*. J. Fish. Assoc. Japan (Nisuishi), 11 (4).

Oka S. 1964. Study of *Penaeus chinensis*. Research Report of Nagasaki University of Fisheries, 17.

Oka S. 1965. Study of *Penaeus chinensis*. Research Report of Nagasaki University of Fisheries, 18.

Oka S. 1967. Study of *P. chinensis* culture, II. Suisan Zoshoku, 15 (2).

Okada T., Kubo I. 1950. Ecological study of Kasumiura freshwater giant prawns. Study of Marine Zoology (Suisandobutsu no Kenkyu), 3.

Saisho K. 1962. Molting and growth of phyllosoma of *Panulirus japonicus*. Fisheries Report of Kagoshima University, 11 (1).

Saishu H. 1967. Relationship of catch to daily vertical migration and bottom water temperature on the basis of periodic estimation of average catch of *Penaeus chinensis*. Seisuikenpo, 34.

Sake, M. 1979. Shrimps—Knowledge and Know-how. Suisansha, Tokyo.

Sheldon D. 1961. Early developmental stages of pink shrimp, *Penaeus duorarum* from Florida waters. Fish. Bull. 190. Fish and Wildlife Service.

Shigeno S. 1947. Number of spawns and eggs of spiny lobster. J. Fish. Assoc. Japan (Nisuishi), 13 (1).

Shigeno S. 1950. Number of spawns of *Panulirus japonicus* V. Siebold. J. Fish. Assoc. Japan (Nisuishi) 15 (11).

Takagi W. 1946. Ecological studies of Nuka shrimp. Biology (Seibutsu), 1.

Takeda F. 1971. *Palaemon* culture. Bull. Hyogo Fish. Exper. Station (Hyogo Suishi Shiken Hokoku), 10.

Ueda T. 1956. Ecology of Japanese inland water shrimps, 1. *Caridina japonica*. J. Zool. (Dobatsuzatsu) 65 (12).

Ueda T. 1957. Ecology of Japanese inland water shrimps, 4. J. Zool. (Dobutsuzatsu) 66 (9).

Ueda T. 1958. Ecology of Japanese inland water shrimps, 5. J. Zool. (Dobutsuzatsu) 67 (5).

Uno Y. 1967. Ecological studies relating to the larvae of *Macrobrachium nipponense* (De Haan). The Sea (Umi), 5 (3).

Uno Y., Kwon, CS. 1969. Larval development of *Macrobrachium rosenbergii* (De Man) reared in the laboratory. J. Tokyo Univ. Fish., 55 (2).

Uno Y., Yagi H. 1980. *Influence de la combinaison des facteurs température et salinité sur la croissance larvaire de Macrobrachium rosenbergii* (De Man) (Palaemonidae, Decapoda, Crustacea). *La mer*, 19 (4).

Wickins JF., Beard TW. 1974. Observations on the breeding and growth of the giant freshwater prawn *Macrobrachium rosenbergii* (De Man) in the laboratory. Aquaculture, 3.

Yasuda J. 1957. Study of biological resources of shrimps in inland bays. Suisangaku Shusei. Tokyo University Publ.

Yoshida T. 1949. Life history of *Penaeus chinensis*. J. Fish. Assoc. Japan (Nisuishi), 15 (5).

3

Shrimp Culture in Japan and Southeast Asia

3.1 SHRIMP SOURCES OF THE WORLD AND CULTURE

Among the so-called "shellfish", crustaceans including shrimp have been important sources as food items throughout the world from time immemorial. Of these crustaceans, shrimp are considered high-grade marine products. The production of crustaceans and the fishing sites according to FAO statistics for 1980 are shown in Fig. 3.1. Production was about 292.8×10^4 tons, of which shrimp constituted 52.1%. Most crustaceans are obtained through fishing and only a small percentage produced by culture. It can be seen from Fig. 3.1 that the fishing sites are not spread out in open seas or oceans but, rather, located in offshore coastal regions. *Kuruma* shrimp, rated as high grade among shrimp species, are collected in tropical and subtropical belts. In recent years the demand for prawns and shrimps from such industrially advanced countries as Japan and the USA has increased markedly. This has resulted in rapid development of the resources in those regions which have become the favorite locations for the natural catch of shrimp. These regions belong to the developing countries for whom the catch can also earn foreign exchange. When the catch is made from a natural resource, a scientific survey for quantity potential is absolutely essential. Yet, in developing countries, there has been a rapid but unplanned increase in catch employing a large number of shrimp trawlers without implementing a survey of the sources before catching operations. As a result of such unplanned catches (with no restrictions set) in some countries, due to the development of coastal regions, the natural sources of some species have been exhausted or greatly reduced. Concomitantly, hasty, likewise unplanned, projects to produce prawns through culturing, since they have a high commercial value, has led to the establishment of ponds by destroying vast acres of mangrove forests, the natural habitats for shrimp during the early stages of their development.

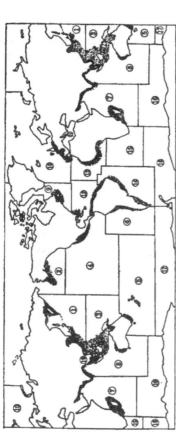

Major crustacean fishing sites worldwide (nos. 1-19 indicate sea regions); ▓ :fishing sites

Sea regions	Crabs A	Shrimps B	Spiny lobsters C	Japanese lobsters D	Krills E	Others F
1	433,474	310,164	1,118	—	—	7,116
2	147,887	43,887	8,126	—	—	—
3	60,618	408,679	4,122	—	—	9,255
4	1,443	79,508	1,851	—	—	2,639
5	238	1,981	4,449	—	—	—
6	5,669	4,284	39	29,243	—	978
7	1,039	200,753	1,354	185	—	14,400
8	7,001	56,389	15,076	—	—	7,942
9	67,555	45,291	37,358	—	—	—
10	36,427	84,218	1,967	41,435	—	643
11	1,384	14,528	872	3,046	—	8,283
12	44,505	171,472	25,005	—	—	785
13	652	33,354	1,986	412	—	4,973
14	12,303	50,200	8,216	—	—	—
15	3,416	976	7,694	—	—	12
16	—	—	—	—	64,312	—
17	—	—	—	—	600	—
18	349	—	—	—	321,344	—
19	—	—	—	—	—	3,326
Total	823,960	1,526,060	111,108	74,321	386,256	6,362

Fig. 3.1. Worldwide commercial distribution and annual production (tons) of crustaceans in 1979 (FAO Statistics, 1980; freshwater production omitted).

This chapter explains the fundamental problems in shrimp culturing and how best to utilize this technique in achieving a rational increment in production of this food item.

3.2   CULTURE AND PRODUCTION IN ASIA

Table 3.1 shows the trend in production through shallow aquaculture in Japan. A gradual increase since 1970 is evident with a leveling off at 100 tons in recent years. This amount constitutes 9.1% of the total production of 11 million tons and is expected to reach 10% in the near future. As obvious from Table 3.1, the range of species is wide and, as explained later, in the countries of Southeast Asia production is high for only certain species. Among the shrimps, mainly production of *Penaeus japonicus* shows a gradual increase. In Japan, in spite of well-advanced techniques in seedling production, the ideal sites for this enterprise have disappeared.

As for Southeast Asia, it can be seen from Table 3.2 that statistical data regarding cultural production is scant. The ratio of cultural production to total catch worked out to 10.2% in 1977. Production of shallow aquaculture, excluding freshwater culture, is given in Table 3.3. The priority production of certain species in some countries vis-a-vis Japan is explained, as mentioned earlier, by the conditions specific to culturing in Japan. For example, milkfish are cultured in the Philippines and Indonesia; cockles in Malaysia and milkfish and oysters in China (Taiwan).

As for *kuruma* shrimp, both Japanese and Taiwanese techniques have been introduced in Taiwan, the Philippines, and Thailand, thereby modernizing traditional cultural practices. In Taiwan, for example, pond culture of *Penaeus monodon*, introduced 10 years ago, is now popular, and the annual production expected to exceed 4,000 tons.

3.3 PRAWNS AND SHRIMPS CULTURED THROUGHOUT THE WORLD

Concomitant with the increase in demand for crustaceans throughout the world, there has been an increase in efforts to obtain higher catches in various countries. *Kuruma* shrimps in particular cannot meet the demand as only a natural resource. Production by well-advanced cultural techniques is now an established industry. In the case of shrimp culture, the fact that it is possible (1) to plan production, (2) to produce a standard quality, and (3) to produce throughout the year and thereby maintain the supply have encouraged producers. These facts have created a boom not only in Southeast Asia, but worldwide.

Table 3.1: Production statistics of shallow aquaculture in Japan (Fisheries Agency of Japan, 1983) (unit: ton)

Year	Total	Fishes								
		Yellowtail	Red sea bream	Crimsontail	Black sea bream	Horse mackerel	Yellowjack	Fugu-rubripes	Leatherfish	Other fishes
1970	549,081	43,300	960	5	—	2	36	26	63	—
1971	608,684	61,743	971	23	—	24	43	21	44	41
1972	647,905	76,913	1,298	95	13	112	15	15	149	113
1973	790,974	80,269	2,606	58	9	348	30	17	253	179
1974	879,761	92,685	3,414	85	4	628	48	8	51	158
1975	772,741	92,352	4,303	126	6	923	22	11	17	236
1976	849,909	101,619	6,453	125	61	721	49	11	10	187
1977	861,389	114,866	8,120	193	67	772	136	18	34	302
1978	917,244	121,728	10,844	532	112	815	181	48	28	720
1979	882,620	154,872	12,253	204	104	1,460	313	74	53	1,228
1980	991,843	149,311	14,757	133	150	2,283	228	69	6	2,780
1981	959,680	150,754	17,953	214	136	3,229	161	163	3	2,291

Year	Invertebrates							Seaweeds			
	Oyster	Scallop	Pearl oyster	Kuruma shrimp	Octopus	Sea squirts	Others	Porphyra tenera (Laver)	Undaria (wakame)	Laminaria (kelp)	Others
1970	190,799	5,675	85	301	109	94	4	231,464	76,698	282	—
1971	193,846	11,165	49	306	98	339	6	244,946	94,350	666	—
1972	217,373	23,162	42	454	68	1,118	42	217,906	105,678	3,338	—
1973	229,899	39,372	34	653	56	4,675	299	311,410	113,158	7,648	—
1974	210,583	62,673	30	911	54	5,036	132	339,314	153,762	10,177	—
1975	201,173	70,256	30	936	41	6,313	114	278,127	102,058	15,696	—
1976	226,286	64,909	34	1,042	42	8,390	89	291,050	126,723	22,087	—
1977	212,786	83,180	39	1,124	16	7,463	94	279,031	125,883	27,249	64
1978	232,069	67,723	38	1,184	11	5,759	216	350,471	102,682	21,890	194
1979	205,509	43,614	40	1,480	22	5,287	178	325,686	103,791	25,291	1,164
1980	261,323	40,399	42	1,546	22	5,749	375	357,672	113,532	38,562	2,904
1981	235,241	59,095	46	1,666	8	6,909	481	340,510	91,272	44,221	5,329

Table 3.2: Fisheries production in countries of Southeast Asia (SEAFDEC, 1981)[1]

Items	Year							
	1972	1973	1974	1975	1976	1977	1978	1979
Total catch (A)	527	552	550	570	604	692	707	709
Cultural production (B)	49	48	51	57	64	71	—[2]	—
Percentage of cultural production (B/A × 100)	9.2	8.6	9.3	10.1	10.6	10.2	—	—

[1] Taiwan, Hong Kong, Indonesia, Malaysia, Singapore, and Thailand.
[2] Data for 1978 not available for Indonesia.

Table 3.3: Cultural production in Southeast Asia in 1978 (SEAFDEC, 1980)

	Malaysia	China (Taiwan)	Hong Kong	Indo-nesia	Singa-pore	Philippines	Thailand
Fishes (marine)							
Grouper	—	—	306	—	—	—	—
Snapper	—	—	103	—	—	—	—
Sea bream	—	—	256	—	—	—	—
Fisheries (estuaries)							
Milkfish	—	29,858	—	48,287	—	118,682	24
Barracuda	—	—	—	571	—	—	—
Mullet	—	251	1,385	3,489	—	—	390
Mollusks (marine)							
Oysters	—	17,994	—	—	—	—	14,594
Limpet	—	—	65	—	—	—	—
Blue sea mussel							49,868
Cockle	55,598	42	—	—	—	—	16,326
Crustaceans (estuaries)							
Kuruma shrimp	—	—	—	15,184	34	—	5,788
Other than *Kuruma* shrimp	—	2,475	—	6,446	—	—	607
Seaweed	—	1,088	—	—	—	—	—
Others	—	31,477	—	10,018	—	—	32,381
Total	55,598	83,135	2,126	83,995	34	118,682	119,978

Table 3.4 lists the species under culture or under experiment. Species cultured commercially include *Penaeus japonicus, P. aztecus, P. duorarum, P. monodon, P. semisulcatus, P. vannamei,* and *M. rosenbergii. P. monodon,* a particularly important species in the tropical region, is mainly cultured nowadays in Taiwan but also in several other countries. Its distribution is wide (Fig. 3.2). The duration of growth from seedling to marketable size is short. It is possible to rear it on feed with a lesser protein content than that required for *P. japonicus.* Further, a large quantity of naturally grown seedlings is available. Given these merits, it will undoubtedly become a popular type throughout the whole of Southeast Asia.

Table 3.4: Types of prawns and shrimps cultured in various countries of the world (Wickins, 1976, partially modified by Uno)

Types	Countries	Types	Countries
Kuruma shrimps		*P. monodon*	Philippines
Metapenaeus affinis	India		Palau
M. bennettae	Australia		India
M. ensis	New Caledonia		Thailand
	Philippines		Mozambique
	Japan		Indonesia
	Taiwan		Malaysia
			Taiwan
M. joyneri	Taiwan	*P. chinensis*	Korea
	Korea		Japan
			China
M. monoceros	India	*P. penicillatus*	Taiwan
	Thailand		China
	South Africa	*P. schmitti*	Cuba
Penaeus aztecus	America	*P. semisulcatus*	Bahrain
	Brazil		Kuwait
	Honduras		Saudi Arabia
			Israel
			Thailand
			Taiwan
P. brasiliensis	Brazil	*P. setiferus*	Costa Rica
			Panama
P. californiensis	Mexico		Honduras
P. duorarum	America	*P. teraoi*	Taiwan
	Honduras		Japan
	Nigeria		
P. esculentus	Australia	*P. vannamei*	Costa Rica
			Panama
			Honduras
P. indicus	South Africa	**Other types**	
	India	*Palaemon serratus*	France
	China		Portugal
P. japonicus	Japan		Spain
	Korea		Brazil
	France		Britain
	Taiwan		
P. kerathurus	France	*Macrobrachium*	Various other
	Greece	*rosenbergii*	countries
	Spain		
	Italy		
P. latisulcatus	Thailand	*Pandalus kessleri*	Japan
	Australia		
P. marginatus	Hawaii		
P. merguiensis	India		
	New Caledonia		
	Indonesia		
	Australia		
	Thailand		

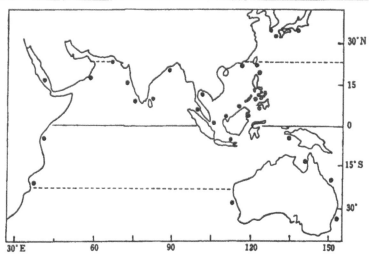

Fig. 3.2. Distribution of *P. monodon* in Southeast Asia and the Indian Ocean (Motoh, 1980).

3.4 BIOLOGICAL CHARACTERISTICS REQUIRED FOR THE CULTURE OF *KURUMA* SHRIMPS

The spawning period of *kuruma* shrimp varies widely from species to species and location of habitat. The physical and chemical conditions of the habitats such as temperature, salinity, natural feed, daylight, etc. vary from place to place and influence shrimp propagation. The spawning period of *kuruma* shrimp in Japan is shown in Fig. 3.3. It can be seen that the mother shrimps used for egg collection are available in March-April in Miyazaki prefecture, May-June in Oiwake prefecture, and July-August in Setonaikai. In other words, the spawning period begins earlier in the southern regions. Ikematsu has given a detailed report on the spawning period of naturally grown shrimp of Ariakekai, which extends from May to October, though there are variations among species. The spawning period begins after the water temperature has risen above 20°C and it was further confirmed that spawning begins when the salinity declines with the advent of the rainy season (Fig. 3.4). There is a close relationship between maturation-spawning of shrimps and temperature-salinity conditions. The embryonic development of the fertilized eggs of *kuruma* shrimp results in hatching of nauplius which lead a planktonic life. Thereafter, undergoing repeated molts they pass through the zoea stage, mysis stage, and postlarval stage and then settle down as bottom dwellers, advancing through the juvenile stage, immature stage, and finally the adult stage respectively. During this period they undergo repeated molts and transformations. The striking feature is that they complete their life cycle by transformation of the body along with change in habitat. This is illustrated in Fig. 3.5. The mother shrimps spawn in June-

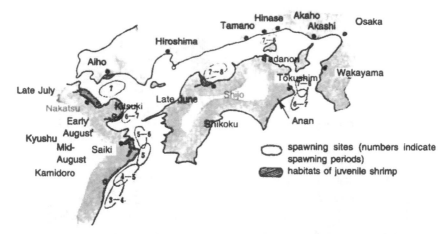

Fig. 3.3. Spawning regions and spawning periods of Japanese *kuruma* shrimp (from the data of Fisheries Agency of Japan).

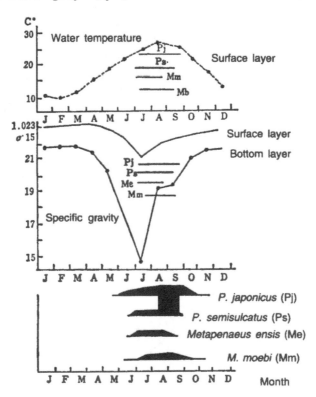

Fig. 3.4. Relationship between spawning period of shrimps of Ariakekai and temperature-salinity (Ikematsu, 1963).

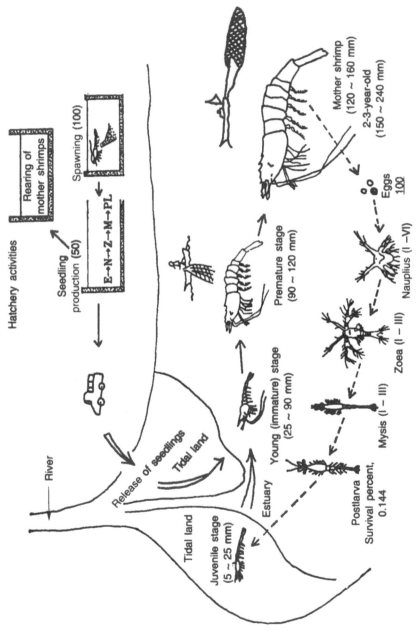

Fig. 3.5. Life history of *kuruma* shrimp (Kurata, 1972, partially modified).

July when the water temperature rises to about 20° C. The eggs are fertilized outside the body and settle to the bottom. At a water temperature of 27-29° C nauplius hatch in 13-14 hours and lead a planktonic life. In *kuruma* shrimp the larvae after hatching undergo 6 molts in 36-37 hours and metamorphose into zoea, which after 3 molts become mysis. The latter become postlarva after 3 molts. Larvae hatching in open seas are carried by water currents and tides into an estuary. There are various approaches to explaining the mechanism whereby planktonic larvae are transported to an estuary. The one most widely accepted is that the reaction to salinity differs between postlarvae and juvenile shrimps and this is the major factor prompting transportation to a river mouth. The appearance of juvenile *Penaeus duorarum* along with tidal movements is shown in Fig. 3.6. The early postlarvae are found in large numbers during high tides and in small numbers during low tides. Contrarily, juvenile shrimps are found in large numbers during low tides and small numbers during high tides. This fact accords with the results of indoor experiments carried out by Hughes (1969). Postlarvae were observed to stop active swimming along the flow direction and to settle on the pond floor when salinity declined. With an increase in salinity, they resumed active swimming along the current direction. In nature, since the swimming strength of the postlarvae is weaker than the tidal current, they are carried by the tide to the estuary. In the case of low tide, they do not sink but move with the tide. Migration to open seas is rare. The results of ecological experiments tally well with practical results. The estuary is rich in small benthic-dwelling organisms that serve as food. This is because organic residue continuously flows down with river waters. For juvenile shrimps which can live hidden in 1-2 cm deep water in tidal land, an estuary is the most suitable habitat. They survive here and grow

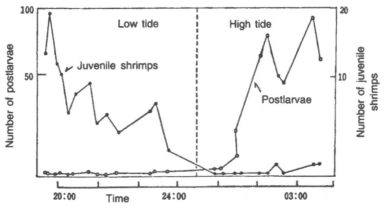

Fig. 3.6. Appearance of *Penaeus duorarum* along with tidal movements. Graphs show variations in collections made every 10 minutes (Hughes, 1967).

into juvenile shrimps that react to salinity in a manner opposite to that of postlarvae. Thus when sea water with low salinity arrives (with low tide), they move along the water flow and migrate to midsea. This is similar to the trophic migration in search of food with high nutrient value observed in marine organisms.

Among the penaeids there are burrowing species such as *P. japonicus*, *P. kerathurus* and others which migrate little during their lifetime and wandering species such as *P. setiferus, M. macleayi* etc. which migrate long distances. In both cases they spawn in midseas, the juvenile forms develop in an estuary or coastal region, after which they return to the midseas, mature, and spawn.

Penaeids are crustaceans which usually live in a wide range of temperature and withstand a wide range of salinity. The survival rates of the green tiger prawn, *kuruma* prawn, and banana prawn against temperature and salinity are shown in Fig. 3.7. All of them have a 100% survival at 8–40° C and 17.5–40.0‰ salinity. According to the experience of culture experts, pond-grown *P. monodon* grows rapidly at a salinity of less than 20% up to about 50 g body weight; at higher salinity production declines. Contrarily, the green tiger prawn grows well even at higher salinity. Therefore in regions of prawn culture sites where the dry and rainy periods are distinct, it is necessary to carefully plan a rational culture by combining the two types on the basis of climatic conditions.

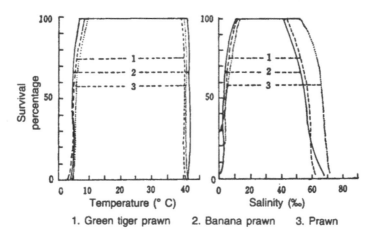

1. Green tiger prawn   2. Banana prawn   3. Prawn

Fig. 3.7. Resistance of early-stage juvenile forms of genus *Penaeus* to temperature and salinity (Motoh, 1982).

3.5 CULTURE METHODS

The culture methods of penaeids are broadly of three types: extensive culture, semi-intensive culture, and intensive culture.

3.5.1 Extensive Culture

In this method the seedlings developed in natural habitats are introduced into the ponds and the adults grown on the feed organisms naturally present in the ponds. It should be noted that no other feed is provided. This method has long been practiced in Singapore, Malaysia, and Thailand. Particularly in Malaysia it is a traditional method with a several-hundred year history. This method uses natural ponds or small inlets of bays. At the time of high tides juvenile forms and even adults enter these ponds. When the sluice gate is closed, the juvenile forms which enter the ponds grow to some extent after 2 to 3 months. They are caught by opening the sluice gate during ebb tide. The sluice gate is made of wood and bricks and the prawns which flow out with the water are netted. The floor of the pond is leveled, a groove dug in one part, and water refilled to a depth suitable for the prawns. In the typical extensive culture method, the prawns grow by feeding on the microorganisms present in the pond. This method is called "trapping pond culture". Although productivity is low, since it requires very little capital, it suits the poorer fishermen. A case example in southern Malaysia is depicted in Fig. 3.8. Culture pond Nos. 1 and 2 are 28.63 ha (with 78.2% effective water area), pond No. 3, 5.6 ha (68.0%), and No. 4, 16.44 ha (38.4%). The annual production is 4,509.6-5,374.8 kg. Details of production are presented in Table 3.5.

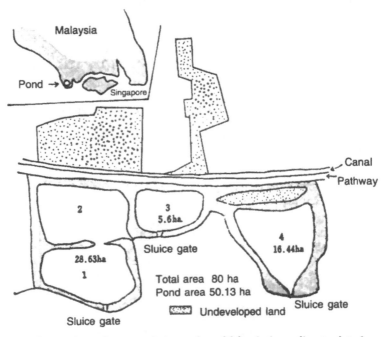

Fig. 3.8. Layout of extensive culture ponds in southern Malaysia (according to data from Hirasawa, 1983).

Table 3.5: Example of production by extensive culture carried out in southern Malaysia (according to Hirasawa, 1983)

Species Items	Annual production		Income	
	1980 (%)	1981 (%)	1980%	1981 (%)
	kg	kg	US$	US$
P. monodon	50.4(1.1)	128.4(2.4)	357.9(2.9)	813.6(5.2)
P. merguiensis	119.6(24.7)	1,231.2(22.9)	6,688.2(53.7)	7,462.1(47.6)
Medium-size prawns (species not known)	332.4(7.3)	328.2(6.1)	1,138.9(9.1)	1,184.3(7.5)
Small prawns (species not known)	1,331.0(29.0)	1,548.6(28.8)	2,786.1(22.4)	4,126.1(26.3)
M. barbata	1,741.2(37.9)	2,138.4(39.8)	1,476.8(11.9)	2,105.4(13.4)
Total	4,527.6(100)	5,374.8(100)	12,447.9(100)	15,691.5(100)

In the case of extensive culture, the results are the outcome of carrying out several types of culture in a mixed form. In this section of sea water it is clear that *P. merguiensis* is the most important species and that *P. monodon* is not rich. Compared to other methods, the productivity of 168.2 kg ha^{-1} is very poor. Catches are taken continuously for a few days every month during the period of high tide. The sluice gate is opened every night during ebb tide and the prawns netted. Fishermen capitalize on the behavior of prawns to actively move about at night in search of food and to rest during the day by hiding in the mud bottom. During high tide the sluice gate is opened at night to allow larvae and juvenile forms from the sea to enter the culture ponds. In this way the prawn resource of the pond is enriched and the culture strength maintained. Details of the catch of *P. merguiensis* during the period of high tide are given in Table 3.6.

Table 3.6: Details of catch from extensive culture ponds in southern Malaysia (according to Hirasawa, 1983)

Date	Number of catches						Total
	1	2	3	4	5	6	
Jan. 31-Feb. 3	18.9	38.1	29.7	24.0	—		110.7
March 5-8	20.4	37.2	46.5	25.2	—		129.3
April 3-6	29.1	27.3	19.8	19.8	—		96.0
May 4-8	26.7	40.8	30.0	27.3	40.5		165.3
June 1-6	21.6	15.3	9.61	14.1	12.0		72.6
July 2-8	28.5	27.9	8.93	12.6	7.8		95.7
Aug. 19-21	16.5	33.3	9.3	31.8	—		120.0
Sept. 1-4	69.6	56.7	22.8	37.2	—		186.0
Sept. 24-30	13.8	20.1	16.5	4.2	16.5	4.8	75.9
Oct. 25-29	11.4	10.8	5.7	6.3	14.1		48.3
Nov. 23-27	10.2	12.5	21.5	24.0	10.5		69.6
Dec. 23-27	9.6	13.2	12.0	15.4	9.6		60.3
Mean	23.0	27.8	22.0	20.2	20.5		

As mentioned earlier, extensive pond culture depends in all aspects on the production capacity in nature. Thus the seedlings, food, water exchange, and other conditions depend on the topographical conditions of the region. Hence it is necessary to use ponds of wide area. In Singapore, it is thought that from the economic point of view, the area should be more than 12 ha. In any case, naturally formed ponds are used. In India, rice fields exceeding 14,000 ha are used for large-scale prawn culture. In West Bengal, large-scale prawn culture is carried out in what are called *bheris* covering a vast area of 9,600 ha in estuaries. By partially introducing semi-intensive or intensive culture techniques it would be possible to improve management and thereby increase production.

3.5.2 Semi-intensive Culture

This is a type of modified extensive culture in which a few artificial techniques are added to the extensive culture. The nutritional requirements of the culture are partially dependent upon the productivity of the environment. This method has been widely popularized in Southeast Asia. Concomitant with technical advancement, efforts are underway in some regions to shift to intensive culture. However, given the relatively low production cost of semi-intensive culture, it is considered economically better suited in Southeast Asian regions. It is currently successfully implemented in Thailand and the Philippines. In the Philippines, natural seedlings are released in ponds of 4-5 ha (0.5-1.5 m water depth). Rice bran and small fishes are provided periodically as supplementary feed for better growth. Inorganic fertilizer is also applied to assist the production of natural food in the ponds by propagating Lab-lab, a combination of diatoms, blue-green algae, and protozoans. As for the structure of the pond, it has a flat bottom with a peripheral groove or canal having a water depth of 0.6-1 m (Fig. 3.9).

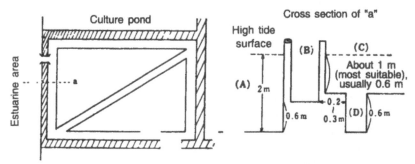

A. Waterway of estuary   B. Main canal for introduction of sea water
C. Pond   D. Peripheral canal

Fig. 3.9. Layout of semi-intensive prawn culture pond.

The peripheral canal inside the pond is very important. The water level of the pond varies considerably depending on full tide and ebb tide. At the time of ebb tide, when the water level lowers, the prawns collect in the canal. This facilitates collection and concomitantly provides a place of escape for the prawns from their enemies. The layout of the canal varies in different countries but the objective is still the same. In the Philippines, single culture of *P. monodon* is carried out in ponds of 4-5 ha (Fig. 3.10, E). Natural seedlings are released (Fig. 3.10, C, D) and, as already mentioned, rice bran, small fishes, etc. are provided at appropriate intervals to supplement otherwise insufficient natural food. Since natural food is essential, to sustain its production inorganic fertilizer is periodically applied to propagate the growth of Lab-lab. The prawn production rate by this method varies greatly, however, depending on the structure and size of the pond, density of seedlings released, conditions of pond bottom, water temperature and salinity, and the culture technique adopted. An example of productivity by the semi-intensive culture method carried out in Panay Island (the Philippines) is presented in Table 3.7. In general, Panay shows low productivity. In the case of *P. monodon*, among the aforesaid conditions, salinity and natural food are extremely

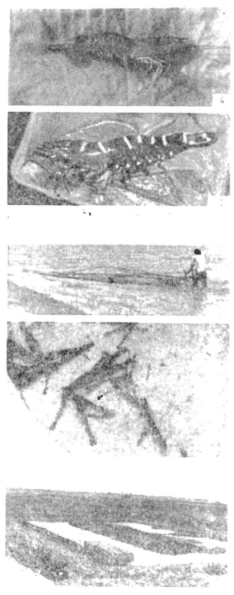

A: Cultured prawns; B: Mother prawn (naturally produced); C: Collection equipment for natural seedlings (fry bulldozer); D: Natural seedlings; E: Pond with high productivity.

Fig. 3.10. Mother prawns, juvenile prawns, collection equipment, natural seedlings, and structure of culture pond of *P. monodon* in the Philippines.

important. A salinity of 20‰ is said to be good and as natural feed Lab-lab appears to be suitable. But when fertilizer is applied, Lab-lab grows excessively and spoils the quality of the bottom water. Moreover Lumut (filamentous algae) tends to cover the pond surface, which hinders prawn growth. To control the growth of Lab-lab and Lumut it is necessary to maintain environmental conditions suitable for prawn growth. For this purpose a mixed culture of milkfish is introduced. At the rate of 100-200 milkfish ha^{-1}, Lumut growth is restricted and Lab-lab growth accelerated. The milkfish feed on Lab-lab proliferating near the pond floor and in doing so the pond water here becomes muddied. The muddied water prevents passage of sunrays and thus growth of Lab-lab is restricted. By this means the excessive growth of natural food is precluded and a suitable quanity of feed maintained. In extensive culture, on the other hand, the major problem even today is the presence of predators of juvenile prawns. When the eggs and juveniles of predatory fishes enter the culture ponds, they develop and feed on juvenile prawns. To minimize such predation water flowing in from the open sea is sieved through a plankton net. Further, concomitant with the release of prawn seedlings, palm leaves are scattered in the pond to provide hiding places for the juvenile prawns. Production conditions for prawn culture in various countries are given in Table 3.8. These days the area priorly used for extensive culture only is being modified for semi-intensive culture, a trend strongly evident in the Philippines.

Table 3.7: Example of pond culture of *P. monodon* on Panay Island (the Philippines) (unpubl. data, Hirasawa, 1983)

Technical level Items	High	Moderate	Low
No. of seedlings ha^{-1}	>5,000	3,000-5,000	Less than 3,000
No. of harvests y^{-1}	>2	1-2	1
Productivity (kg per harvest per ha)	>240	About 120	<50
Survival %age	>60	40-60	<40
Supplementary diet	Plants and animals	Mostly plants with few animals	Absent
Sluice gate	2	1-2	1
Water depth (m)	>1	1	<1
Bottom quality	Leveled, with canals	With root scars of mangrove and canals	Root scars of mangrove and no canals
Mixed culture	Absent	Milkfish	A few milkfish

Table 3.8: Extensive and semi-intensive culture in Southeast Asia (according to Tham, 1968; Pedini, 1981)

Items Countries	Types	Culture method	Prod. (kg ha^{-1} y^{-1})	Pond (ha)	Water change
Thailand	P. merguiensis P. monodon P. indicus M. brevicornis M. ensis	Extensive culture	25-625	10(2-80)	Pump or tidal currents
Philippines	P. monodon P. merguiensis Prawns of genus Metapenaeus	Semi-inten- sive culture	100-300	4-5	Tidal currents
Indonesia	P. monodon P. merguiensis P. indicus	Extensive	100-300	2-3	Tidal currents
Singapore and Malaysia	P. merguiensis P. indicus Metapenaeus sp.	Extensive	837[1]	4-50	Tidal currents
India	P. merguiensis P. indicus Metapenaeus sp.	Extensive	500-1,200	4,500-14,000[2]	Tidal currents

[1] Example of pond with high productivity in Johor, S. Malaysia.
[2] Carried out in backwaters with wide paddy fields and complex water system.

Supplementary feed, exclusion of predators, increased feed production by periodic application of fertilizer, controlling pond environmental conditions through a mixed fish culture, and ensuring water change by means of a pump, have markedly raised productivity. In the Pulakan region of the Philippines, ponds which in the past were used for milkfish culture have now been converted to the culture of giant tiger prawn which has a high economic rating. Its productivity has increased to 1-2 tons ha^{-1} y^{-1}.

3.5.3 Intensive Culture

Developed in Japan and also popular in Taiwan. In Kagoshima Prefecture of Japan it has become possible to produce 2-2.5 kg m^2 of *kuruma* shrimps by this technique. As mentioned earlier, this is a cultural method wherein artificial techniques are adopted from seedling production to the adult stage. This method is described below under three aspects: seedling production, intermediate rearing, and advance rearing.

(1) Seedling production

For the collection of eggs mostly naturally grown mature prawns are used.

The mature prawns to be used for egg collection are selected from those caught during the spawning period. The ovaries in these are in a ripe state. Figure 3.11 shows the course of development of ovaries in *Penaeus monodon*. From stage B of the Figure the ovaries start to yellow and in stages C-D change to bluish-pink. It is possible to observe the developmental changes in the ovaries externally. The prawns close to stage D are selected and released into the spawning tanks. According to Motoh (1981), spawning occurs at night. The mother prawns resting on the bottom of the spawning tanks suddenly begin to swim vigorously at the time of spawning. They calm down within a minute or two and swim gently when they start spawning. While spawning they alternately clasp and open the 2nd-5th walking legs. The ova are released through the genital apertures at the proximal part of the 3rd walking legs and the sperm are released from the thelycum at the proximal part of the 5th walking legs. During this period they rapidly "pedal" the swimming appendages. This behavior stirs the sea water and brings about fertilization. Spawning behavior lasts about two minutes but is repeated any number of times. If there is no camera flash they continue to spawn. Upon completion of spawning the mother prawns settle down to the tank bottom and unlike *P. setiferus*, do not die immediately after spawning.

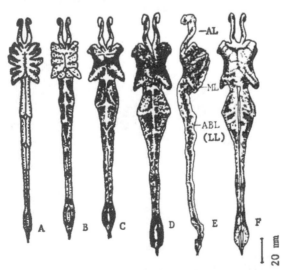

A: Undeveloped or after egg release (dorsal view); B: During course of development (dorsal view): C: Somewhat matured (dorsal view); D: Matured stage (dorsal view); E: Lateral view of D; F: Ventral view of D. AL: anterior lobe; ML: median lobe; ABL: ventral lobe (or LL: lateral lobe)

Fig. 3.11. Development of ovary in *Penaeus monodon* (Motoh, 1981).

The fertilized eggs hatch as nauplius after about 12 hours at 26-29° C water temperature and 29.5-34.2‰ salinity. The nauplius molt 5 times in 1.5 days and pass through 6 substages to attain the zoea stage. Zoea grow into mysis after molting 3 times in about 5 days. Thereafter mysis undergo 3 molts in 6-15 days and grow into postlarvae. Thus after hatching the larvae become postlarvae in 12.5-21.5 days. For 4-5 days after attaining the postlarval stage they are reared in the larvae rearing tank up to the P_{4-5} stage. This is known as intermediate rearing and is explained below. They are grown to marketable size. The transformation from egg to postlarva varies depending on the species of prawn, for example, *kuruma* prawn, *P. japonicus*, *P. chinense*, *P. merguiensis* etc. However, the course of change from nauplius to zoea to mysis and to postlarva is the same. Figure 3.12 shows the morphological changes in *P. japonicus* during the larval stage. In *P. japonicus*, the serrations on the maxillipeds of the larvae are well developed while in *Macrobrachium* serrations are

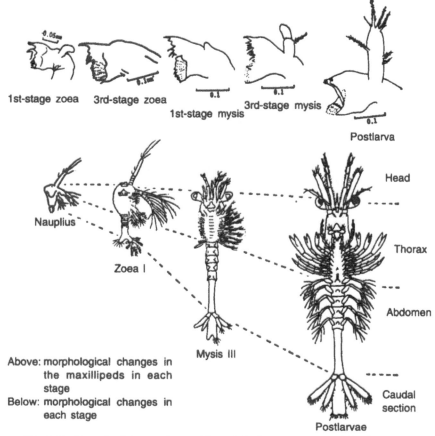

Fig. 3.12. Growth of *Penaeus* larvae (data of Hudinaga, 1942, modified by author).

almost lacking. This indicates that compared to *Macrobrachium*, *Penaeus* is better adapted to feeding on plants. That is, it has a strong tendency to be phytophagous. It is well known that unicellular phytoplankton are effective as food for the early stage larvae of *Penaeus*. Table 3.9 shows the rearing methods of larvae currently in practice, and in particular the common method adopted in Japan. The important points to be noted are that to obtain eggs of good quality, it is necessary to preserve mature prawns and to make arrangements beforehand for supplementary feed. In the supplementary feed system, as explained earlier, the phytoplankton are maintained at 10,000 cells cc^{-1} before the larvae grow into zoea and start to feed. As the larvae grow, the dosage should be increased to about 50,000 cells cc^{-1}. When the mysis stage is attained, it is desirable to add feed of higher nutritive value such as *Artemia* larvae, fish meat, small pieces of bivalves, etc. In large-scale production, tanks with a capacity of 200 tons (10 × 10 × 2 m) are commonly used. In these tanks 800,000–1,250,000 of P_{15-25} larvae are produced. Just recently an artificial diet was developed for growing larvae, which has greatly facilitated rearing from postlarva to P_{20}. Using this artificial diet, private prawn producers as well as cooperative fishing agencies are able to produce as many as 10 million seedlings. Technical standards of seedling production vary notably from country to country and also depend on the species. The actual conditions are given in Table 3.10. In this Table, the production technique of *P. merguiensis* in particular is of a high grade. The conditions for larval rearing by the technique developed in China, according to

Table 3.9: Seedling production of *Penaeus*

Develop-mental stage	Mother prawn	Embryo	Nauplius	Zoea	Mysis	Postlarvae	Seedling P_{20}
Period (days)		1-3	0.7	2	5-7	3	20-25
Body length (mm)	160-220	0.3	0.3-0.5	0.9-2.5	2.8-4.0	5-7	11-13
Feed	Fertilizer for phytoplankton $\begin{cases} KNO_3 : 1 \text{ g/m}^3 \\ Na_2HPO_4: 0.1 \text{ g/m}^3 \\ Na_2SiO_3: 0.05 \text{ g/m}^3 \end{cases}$			Diatoms Rotifers Artemia		Clams Mysids Misc. shrimps Artificial diet	
Water depth				10-12 days raised daily by 10 cm		Water changed daily at 1/3-1/5	
Number of prawns m^3	(Spawning 20%; eggs released 200,000; eggs fertilized 40,000)			30,000 (0.7 hatching %)		20,000-25,000	10,000-15,000
Survival (%)	100		70-80	50-60			25-40

Table 3.10: Seedling production of *Penaeus*

Items / Types	Size (mm) Egg	Size (mm) Post-larvae	No. days to post-larvae	Post-larval density (larvae/ton)	Temp. °C	Remarks
P. chinensis	0.13-0.14	2.3	17-20	42,000	20-24	Size of egg and post-larvae from Oka, 1967; rest from Wang *et al.*, 1981
P. japonicus	0.26-0.28	4.79-5.0	16	6,000-25,000	26-28	Size of egg and postlarvae from Hudinaga, 1942
P. merguiensis	0.27	5.13	10	4,800-10,700	26.3-31.8	No. days to postlarvae and density from Terazaki, 1981; rest from Motoh, 1979
P. monodon	0.29	5.74	12-21	10,000-16,000*	26.9-29.0	Motoh, 1979, 1981

*Results obtained at rearing site located in eastern bay of Taiwan.

Wang *et al.*, (1982) are shown in Fig. 3.13. From the second stage of zoea, the larvae are grown on soybean milk. The spawning tank ($1.2 \times 0.8 \times 1.0$ m) is used for larval rearing. After spawning the water tank is cleaned twice. Nauplius from stage 1 to stage 7 are released ($10\text{-}15 \times 10^4$ m^3). As shown in the Figure, feed is provided along with proper aeration. The rearing density is very high. This is because the larval rearing technique is excellent and concomitantly the larvae very small in size (Table 3.10). Considering the fact that the proper water temperature is low, it appears that rearing is much better in this case than in other species.

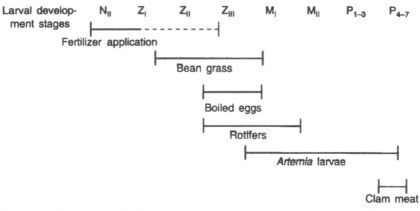

Rearing conditions: light 30,000-40,000 lux; tank, 100-ton capacity (water depth 1.4-1.5 m); water temperature 2-3° C and higher than room temperature; aeration 20-40% month^{-1} (water exchange); same, 40-50% in later stage; rearing density: $10\text{-}15 \times 10$ Z/m^3; survival 17.4%.

Fig. 3.13. Feeding system in high-density rearing of larvae of *P. chinensis* in Yamato (China) (Wang *et al.*, 1982).

(2) Intermediate and advanced rearing

As explained above, the postlarvae of P_4-P_{20} reared indoors are then transferred to small open-air ponds and allowed to grow to a size of 3 cm. The intention of intermediate rearing is to prepare seedlings suitable for release. In Japan, the larvae are reared indoors up to P_{20} (0.01 g). They are later transferred to open-air ponds and allowed to grow to the juvenile stage. During the course of growth of postlarvae, they are suitably divided into sections and intensive rearing is carried out. An example of such rearing is shown in Fig. 3.14 and Table 3.11. When quantitative production is targeted, the sea water used for culturing is filtered through a plankton net to prevent entry of fingerlings and eggs of predatory organisms such as sea bream, perch, gobby, etc. At the same time, it is necessary to make a plan which will ensure uniform prawn growth; for this purpose, the juvenile prawns grown from the intermediate culture in sections are separated out. In various countries of Southeast Asia, in particular Taiwan, continuous harvest, that is, large-scale production is often done. In that case the individuals grown to marketable size are carefully selected and packed. In Japan, depending on topographical conditions, five different types of cultural methods are adopted. Each has its special features as shown in Table 3.12. For example, in the inner bay of Seto (Setonakai), semi-intensive culture is carried out by utilizing the salt fields close to the culture ponds. On the other hand, at Kagosima, large circular tanks set on land are used for high density culture. Between these two extreme types, there are three others. In order to establish prawn culture, it is essential to consider the size of the pond, scale of operation, electric power, cultural technique standard, and other factors. Concomitantly, proper analysis of the economy should also be done and the method best suited for the region then determined.

The production rate of *kuruma* shrimp culture of Japan was 1,042 tons in 1976. In 1981, it rose to 2,235 tons, certainly a rapid increase within a short period. However, at present the water areas suitable for *kuruma* shrimp culture have reduced and therefore further increase in production cannot be expected. As against this picture, the seedling production in 1980 had rapidly increased to about 6 trillion. These seedlings were released into natural waters, thereby increasing this resource in the natural environment. This maintains the fisheries catch. Experiments are presently being carried out in various regions to obtain a fisheries catch based on resource culture through the release of seedlings. In Shizuoka prefecture, the seedlings produced by the National Center for Fisheries Culture were released in Lake Hamana (the seedlings of P_{35}-P_{40} measuring 12.4–16.1 mm body length, amounting to 6,682,000 individuals, were cultured in an area of 1,600–6,700 m^2 covered by a net). The density of culture in 1980 was 170-330 PL m^2. The culture was continued for 2 weeks during

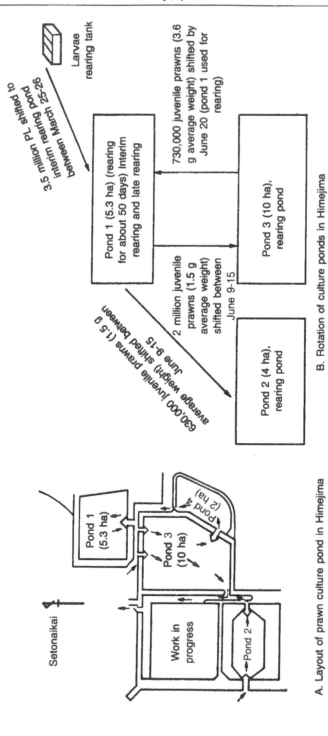

A. Layout of prawn culture pond in Himejima

B. Rotation of culture ponds in Himejima, Oiwake prefecture (Kurata et al., 1976).

Fig. 3.14. Prawn culture in Himejima

Table 3.11: Prawn production in Himejima, Oiwake prefecture (Kurata *et al.*, 1976)

Pond no.	No. released	Catch		Average body weight	Feed supply (ton)	Survival	Feed efficiency
		Number	Tons				
1	730,000	788,400	14.08	17.8	93.54	108	6.6
2	630,000	636,300	11.43	17.9	68.61	101	6.0
3	1,050,000	997,500	14.03	14.1	137.80	95	9.8

Table 3.12: Special features of five types of *kuruma* shrimp culture methods (Hirasawa *et al.*, 1967)

Items Pond	Type of pond	Pond area (ha)	Water exchange (%/day)	Prod. (g m²)	Feed effic.	Culture method
Setonaikai						
Type A	Improved salt field	1-5	25	250-300	10-14:1	Semi-intensive
Type B		0.5-1	50	500-600	10:14:1	
Tenso						
Type C	Net covered	0.5-1	90	250-350	12-15:1	Semi-intensive
Kago-shima						
Type D	Circular tank on land	0.1	2.0	1,800-2,000	22-22:1*	Intensive
Type E		0.1	4.0	2,700-3,000	17-20:1*	

*Dehydrated artificial diet was also included at the rate of 6 parts raw to 1 part dehydrated.

which the larvae grew to 3 cm. It was possible to obtain 25.3-49.1% juvenile prawns. These were released by removing the net earlier used in the intermediate culture. The yield in this case was 13.0%.

Resource culture fishing involves increasing the resource by releasing the seedlings and catching the fully grown individuals. It is an economically profitable fishing operation but can only be successfully carried out in certain regions. There are great expectations to develop resource culture prawn fishing by using such wide areas as the mangrove belts in Southeast Asia.

REFERENCES

Hamanako Branch of Fisheries Experiment Station, Shiru Oka Prefecture. 1981. Report on technical development of *kuruma* shrimp release in 1980, 221: 1-48.

Hirasawa Y, Walford J. 1976. The economics of *kurumaebi* (*Penaeus japonicus*) shrimp farming. FAO, FIR: AQ/Conf./76/R., 27: 1-19.

Hudinaga M. 1942. Reproduction, development and rearing of *Penaeus japonicus* Bate. Jap. J. Zool., **10**: 305-393.

Hughes DA. 1967. On the mechanisms underlying tide associated movements of *Penaeus duorarum*. Contr. to FAO World Scientific Conf. on the biology and culture of shrimps and prawns, Mexico City.

Hughes DA. 1969. Responses to salinity change as a tidal transport mechanism of pink shrimp, *Penaeus duorarum*. Biol. Bull., **136**: 43-53.

Ikematsu T. 1963. Life history of prawns and mysids in Ariakekai. Study relating to ecology. Seikaiku Suikenpo, **30**: 1-24.

Kurata H. 1972. Certain principles pertaining to the penaeid shrimp seedling and seedling farming in the sea. Bull. Nansei Reg. Fish. Res. Lab., **5**: 33-75 (in Japanese with English abstract).

Kurata H., Shigeno K. 1976. Recent progress in the farming of penaeid shrimp. FAO, FIR: AQ/Conf./76/R. **17**: 1-23.

Motoh H. 1979. Larvae of Decapoda, Crustacea of the Philippines. III. Larval development of the giant tiger prawn, *Penaeus monodon* reared in the laboratory. Bull. Japan. Soc. Sci. Fish., **45**(10): 1201-1216.

Motoh H. 1981. Studies on the fisheries biology of the giant tiger prawn, *Penaeus monodon* in the Philippines. SEAFDEC AQD. Tech. Rept., **7**: 1-138.

Namotsu-sho Statistics Publ. Division. 1979. Annual Report of Fisheries Culture Production Statistics. Association of Agriculture and Forestry Statistics, pp. 1-317.

Oka S. 1967. Study of the culture of *P. chinensis*, II. Seedling production and culture. Suizanzoshoku, **15**(2): 7-32.

Pedini M. 1981. Penaeid shrimp culture in tropical developing countries. FAO Fish. Circular, **732**: 1-14.

SEAFDEC. 1981. Fishery Statistical Bulletin for the South China Sea Area, 1979, pp. 1-201.

Terazaki M. 1981. Mass production of the young banana prawn, *Penaeus merguiensis* De Man. La mer, Tokyo, **19**: 23-29.

Tham AK. 1968. Prawn culture in Singapore. FAO Fish Rept., **2** (57): 85-73.

Uno Y, Hirasawa Y, Walford J, Hiramatsu K. 1983. Report for Year One of the research programme: Assessment of the present levels of aquaculture techniques and management in coastal zones of Southeast Asia, pp. 1-87.

Wang K., Li D., Meng Q., Gao J., Li S., Yu K., Lu T. 1982. Studies on techniques for industrialized fry production of *Penaeus orientalis* Hishinouye. J. Shandong College of Oceanology, **12** (3): 65-71.

Wickins JF. 1976. Prawn biology and culture. Oceanogr. Mar. Biol. Ann. Rev., **14**: 435-507.

4

Prawn Culture in Central and South America

4.1 CULTURE CONDITIONS IN CENTRAL AND SOUTH AMERICA

4.1.1 Geographical and Biological Environment

Central and South America comprise more than 30 countries with extremely diverse topography (relief). These countries include Mexico, the southern boundary of North America, Cuba and Trinidad in the Caribbean Sea, Boliva and Paraguay which have no coastline and, on the other hand, Chile and Argentina with extensive coastlines reaching almost to the southern tip of the South American continent. Of these countries, more than 10 have coasts rich in natural prawn/shrimp resources. For example, the Mexican Fishing Center, Guyana Fishing Center, Ecuador Fishing Center, and others are well known for shrimp fishing. Among the top ten countries with a production exceeding 30,000 tons aside from North America, are Mexico and Brazil. Others in the list are India and countries of Southeast Asia. Several countries of Central and South America, in particular those located in the tropical belt, have the environmental conditions requisite in terms of topography, water quality, etc. for prawn production.

Table 4.1 shows the types of large prawns living in Central and South American waters and their size. Except for three, all the types measure more than 20 cm in body length.

Table 4.1: Types and maximum body length of large-size *Penaeus* caught in Central and South America (from Holthuis, 1980)

Pacific coast		Gulf of Mexico, Caribbean Sea	
Types	Total length (mm)	Types	Total length (mm)
Penaeus californiensis	209 (♀)	*Penaeus aztecus*	236 (♀)
Penaeus stylirostris	230	*Penaeus setiferus*	200 (♀)
Penaeus vannamei	230	*Penaeus duorarum*	280 (♀)
Penaeus occidentalis	215	*Penaeus schmitti*	235 (♀)
Penaeus brevirostris	170 (♀)	*Penaeus substilis*	205 (♀)
		Penaeus notialis	192 (♀)
		Penaeus brasiliensis	191 (♂)
		Penaeus paulensis	215 (♀)

There are only 9 species in the world which as a species per se yield more than 10,000 tons annually. Of these, 4 species, viz., *Penaeus aztecus* (40,000 tons), *P. setiferus* (30,000 tons), *P. duorarum* (20,000 tons), and *P. californiensis* (20,000 tons), live in these regions. This signifies also that prawn productivity is quite high in these coastal areas.

4.1.2 Socioeconomic Milieu

The countries of Central and South America are located in the Southern Hemisphere and the resources they possess are supplied as raw materials and primary products to countries in the north. In this way they maintain their economic balance. Nowadays, even to produce the primary products in a consolidated manner, vast capital is required. Thus with Mexico and Brazil leading the list, most of the countries are desperately dependent on foreign exchange from the northern countries. The authorities in every country are well aware that the limited resources do not allow rapid development of international trade. They are also aware that in order to sustain the resources and divert them toward an industrial society, there is an urgent need to establish a stable society and concomitantly accumulate vast capital and successful techniques. For this purpose practical efforts are directed toward financial assistance and tax measures suitable for the producers of international trade items, including fisheries, devising a technical education program, and also setting up cooperative agencies.

4.1.3 Practical Conditions of Prawn Culture

Production of prawns by culturing began as early as 1960 along the southern coasts of Ecuador during the course of natural production. In early 1970, culturing was undertaken on an experimental basis in the Honduras, Venezuela, Brazil, and Peru.

The desire to invest in prawn culturing strengthened in 1970 when the world production of prawn by fishing became restricted, the cost of production increased, and the demand as well as price increased. Prawn culture in an organized industrial form was put into practice in 1977. However, even if it is assumed that many of the South American countries have ideal culturing sites along their coasts and an ideal environment with suitable species, much capital is needed to develop the culture on an industrial scale. Moreover, yield should be assured and the operation stable. But fundamental differences in every society and in economic conditions pose major constraints. The commercial cultural activities in various countries in 1983 are shown in Fig. 4.1. Only 9 countries have practicable cultural schemes. Of these, again only 4, namely, Costa Rica, Panama, Ecuador, and Peru along the Pacific coast are on a commercial profit footing. Other countries are carrying out culture on a trial-and-error basis.

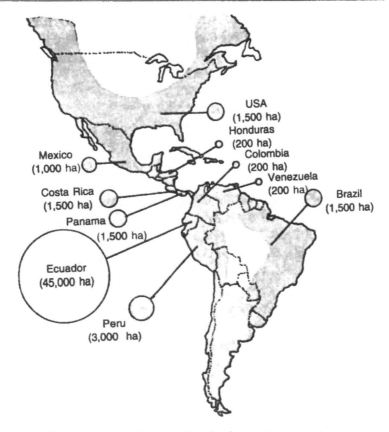

Fig. 4.1. Distribution of culture ponds in South American countries.

The conditions of various countries shown in Fig. 4.1 are described below.

(1) Mexico

Following India and the USA, Mexico holds third position in terms of production. It is possible to develop 60,000 ha of culture grounds all along the coast. In 1977, Coca Cola Company planned to establish culture over an area of 10,000 ha but due to unmitigating pressure from prawn fishing groups and certain social restrictions, the proposal had to be dropped.

The importance of increasing production through the culture method is recognized by Mexico and for this purpose experimental ponds have been established in several sites over an area of about 1,000 ha. However, production in relation to the area is extremely poor. The annual production of cultured prawns is estimated to be 200 tons.

(2) Costa Rica

In Central and South America, Costa Rica is said to be socially the most stable country. For many years now heavy capital investment has been forthcoming from the USA and Europe. In 1977, aquaculture was promoted by Pepsi Cola and two or three other private concerns. In 1979, a company called Maricultura received international capital of 10 million dollars and established hatcheries and culture ponds on 400 ha. At present, the total area of the culture ponds is said to be 1,500 ha. Since production maintenance follows a positive approach to feed application, productivity is high and the annual production of cultured prawn is purportedly 1,500 tons.

This production rate has completely surpassed production through fishing. Such a reversal is also found in Ecuador and Peru.

(3) Panama

As in the case of Costa Rica, experimental ponds of 350 ha have been established by the Ralston Purina Company and Coca Cola of America. Besides, several local private concerns have also set up culture ponds. At present the total area covered is about 1,500 ha. The production pattern here, as in Costa Rica, follows culture by supplementary feeding. The total production is reportedly close to 1,500 tons.

(4) Ecuador and Peru

These are discussed in Section 4.2.

(5) Honduras

Since 1973 research organizations have been carrying out experimental culture of 7 local species of prawn in open-air ponds. Had the approach been good, much progress would have been achieved. However, owing to lack of proper communication concerning current activities, the total area of culture ponds was restricted to 200 ha and annual production has not exceeded 100 tons.

(6) Colombia

With a coastline bordering both the Pacific Ocean and the Caribbean Sea, there are vast areas suitable for cultural activities but the country has not drawn up positive measures for such development. Further, infrastructure in the coastal region is almost nil. Experimental ponds of about 200 ha are said to be located along the northernmost part of the Caribbean coast. Production is presumably very low.

(7) Venezuela

In 1973, hatcheries and culture ponds were tried in experimental centers. Since the capitalists of Venezuela concentrated mostly on the petroleum industry, hardly any investment was made in aquaculture. The pond area is less than 200 ha and production is presumably very low.

(8) Brazil

The country is well known for its fishing ground in the Amazon River estuary in the north and the Santos-Riogrande fishing ground in the south. As a prawn producer it holds 6th place in the world. It has locations suitable for excellent culture in the Amazon River estuary and in the northeast region. Under proper conditions, it has enormous potential for rapid development but presently the total area is limited to about 1,500 ha. Productivity varies widely depending on the entrepreneur. Total production is said to be about 500 tons. Experimental culture was initiated early in Brazil, commencing in the northeast region in 1972. In 1974, private enterprises in Rio initiated hatching and rearing activities.

On summarizing the pond areas and production of the various countries surveyed above, it is found that at present the total area of culture ponds in Central and South America is 54,000 ha and the total production from these ponds 22,000 tons. Annual productivity per ha is 400 kg. The total quantity of tropical prawn obtained by fishing from these waters is 300,000 tons. Thus the productivity of cultured prawns exceeds by 7% the quantity obtained by fishing. This percentage is expected to increase further.

4.2   CULTURE CONDITIONS IN ECUADOR AND PERU

4.2.1   Development of Culture

(1) Ecuador

The coastal areas in the northern part and those in the Gulf of Guayaquil in the southern part are covered with mangroves and have many estuaries. As these are ideal rearing places for juvenile prawns, prawns live along the entire coastline. Hence prawn fishing has long been carried out in the so-called Ecuadorian fishing ground. The recent increase in price of prawns has accelerated positive developments in Ecuador and prawn export has been established as an important core industry.

In the past Ecuador was a monoculture agricultural country producing banana, coffee, cocoa, etc. However, when petroleum was developed and the fishing boats built stronger and modernized, the chief industry of 1982 was petroleum, followed by marine products. Among the marine products, prawn production constituted a very high percentage. In the past, the annual prawn production by fishing reached 5,000-8,000 tons. However, even by increasing the number of fishing boats (about 280 in 1977), since the productivity of the fishing ground is limited, not only did production fail to register an increase, there was a decreasing trend in the profit realized by various entrepreneurs. So interest in culture was revived and investments were made in this field.

Prawn culture of Ecuador was started in 1960 by farmers of the coastal region in the southern part of the Gulf of Guayaquil. In 1969, culture ponds were set up on a commercial basis. However, the so-called "gold rush" of prawn culture began in 1977. Prawn culture progressed rapidly through the consolidated efforts of agriculture capitalists owning land in the coastal region and the earlier mentioned fishery capitalists.

A private culture enterprise called El Rosario of Guayaquil has represented the progress made over a 10-year period from 1977 in figures (the figures for seedlings and feed after 1982 are assumed). Table 4.2 outlines the progress from 1977 to 1981.

Table 4.2: Changing trends in the culture pond area and production in Ecuador (according to the El Rosario Company)

Year	Pond area (ha)	Production (t)	Productivity (headless) kg ha^{-1}
1977	3,000	818	258
1978	5,500	1,682	306
1979	8,200	2,545	310
1980	18,570	5,909	318
1981	27,000	9,091	336
1982 (assumed)	35,000	12,727	380
1983 (assumed)	45,000	17,818	396
1984 (assumed)	55,000	24,955	454
1985 (assumed)	62,000	34,955	564
1986 (assumed)	65,000	48,909	752

According to El Rosario, the annual prawn production through fishing in 1982 was 3,200 tons. The annual prawn production through culture in 1980 had already surpassed the production from the seas and in 1982 was expected to reach 398%. At present the prawn production of Ecuador is completely dependent upon culture.

The aforesaid values are graphically depicted in Fig. 4.2.

As per data for 1983, there are three countries in the world, namely, Taiwan, India, and Ecuador, which produce more than 10,000 tons (annually) by the culture method. If as shown in Table 4.2 production attains 18,000 tons, Ecuador will occupy first position in the world and prawn producers worldwide will naturally take notice of this fact.

(2) Peru

Tumbes region at the northernmost edge of Peru close to Ecuador includes the northern half of the coast which constitutes the southernmost part of the Gulf of Guayaquil. At its juncture with the Ecuadorian prawn fishing ground it has an extremely narrow range of the same. For some years now prawn fishing has been carried out by a single private concern. With 20-30 fishing boats, 300-500 tons of prawns are caught annually. However, the main item of the marine industry in Peru is anchoveta.

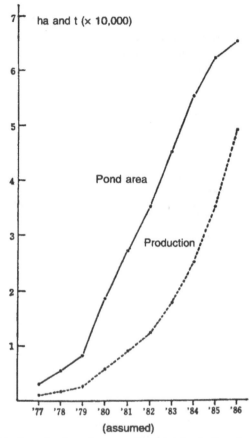

Fig. 4.2. Growth of pond area and production in Ecuador.

Up to 1970, Peru topped the world in anchoveta catch. It acquired the status of a country with maximum supply of this fish. Hence it showed little interest in its prawn resource. Unfortunately, owing to erratic catch and abnormal weather conditions (El Nino) the anchoveta fishing industry rapidly deteriorated and production in 1977 was about half that in 1970. A present, Peru has been pushed to 7th position in the world.

A survey of prawn culture in Peru was initiated in 1971 by the government. Commercial-size experimental ponds were set up in 1976 and the commercial prospects were recognized. At the end of 1977, 6,300 ha along the Tumbes coast were established as a public sector enterprise and culture ponds developed. After obtaining sanction from the government several enterprises set up culture ponds and from 1978 cultural activities were initiated. This was the positive start of prawn culture in this country.

Unlike Ecuador where prawn culture started as a more or less natural development and has accumulated almost 20 years of experience, in Peru it started from scratch. Five years after its start, i.e., in 1983, the pond area was widened to 2,500 ha and the production of headless prawns is said to reach 1,250 tons. Thus it has markedly surpassed marine production.

Table 4.3 gives data relating to cultural activities of Peru compiled on the basis of surveys carried out from 1978 to 1981, the period when the author lived in Peru. It shows the systematic progress from the starting point, taken as zero.

Table 4.3: Changing trends in culture pond area and production in Peru

Year	Pond area (ha)	Production (t)	Productivity (headless) (kg ha^{-1})
1977	0	0	0
1978	310	20	65
1979	710	130	183
1980	1,225	325	265
1981	1,865	775	416
1982 (assumed)	2,245	1,070	477
1983 (assumed)	2,500	1,250	500
1984 (assumed)	2,600	1,340	515
1985 (assumed)	2,800	1,470	525
1986 (assumed)	3,000	1,650	550

Figure 4.3 graphically depicts the values given in Table 4.3. It shows the variations occurring every year between growth of pond area and growth of production.

The progress of culture in Ecuador and Peru have been discussed. Figure 4.4 gives the sites of ponds developed along the coastal regions of the two countries. It is evident that more than 80% is concentrated in the coastal region of the Gulf of Guayaquil.

In Ecuador, culture was first initiated in El Oro district in the southern part of the Gulf of Guayaquil and the culture area covered 13,000 ha (39% of the country). Guayas district in the northern part of the Gulf covers 14,000 ha (43%). At present, this region has the maximum concentration of ponds. This is because grounds suitable for culture are widely distributed here. Moreover, Guayaquil is the largest city in Ecuador and has the best port establishment. Furthermore, most of the frozen-food processors are based here. In addition, 5,400 ha (16%) in Manabi district on the northern coast and 600 ha (2%) in Esmeraldas district near the Colombian border have been developed. At the end of 1982, 112 private enterprises were using these ponds for culture.

In Peru, the ground suitable for culture is confined to Tumbes district and the maximum area is limited to 6,300 ha. In 1983, ponds were

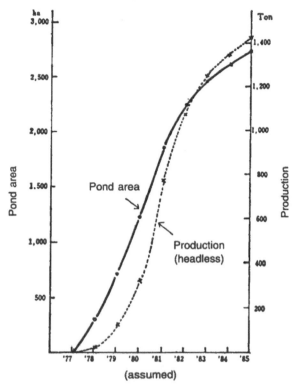

Fig. 4.3. Growth of pond area and production of Peru.

developed on just half of the permitted area and 24 private enterprises were using them in the 3,000 ha area.

4.2.2 Culture Method

(1) Types and characteristics of cultured prawns

As shown in Table 4.1, five species of the genus *Penaeus* live along the Peruvian coast. Figure 4.5 shows the morphology of these species. Of them, only 2, viz. *P. vannamei* and *P. stylirostris*, are suitable for culture production; the other 3 are not. Natural seedlings of the first two species are readily available on a large scale and, moreover, pond yield is very high. Seedlings of *P. occidentalis* are also periodically caught in large numbers but their survival in rearing ponds is almost nil. Regarding the other two species, *P. californiensis* and *P. brevirostris*, very few seedlings enter the estuaries as they mostly dwell away from the coast. Hence they are not practicable for culture.

The first two species of prawns are ideally suited for culture. Seedlings are caught in large numbers and live throughout the year in the

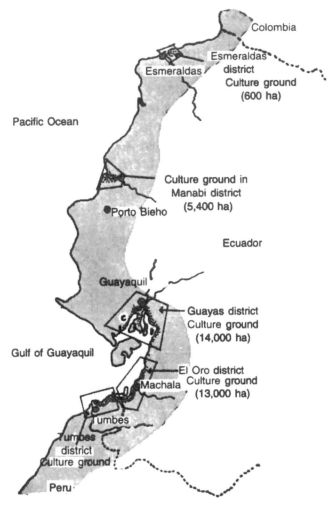

Fig. 4.4. Distribution of culture ponds in Ecuador and Peru (FOFA, 1982).

coastal mangrove region. Moreover, they are highly resistant to environmental changes. For example, they can withstand salinity of less than 10‰ to more than 70‰ and a temperature of almost 40°C. Since they are not in the habit of hiding themselves, they can resist mud bottom blackening due to high hydrogen sulfide content. Hence they have a high rate of survival even under adverse environmental conditions. Among the prawns, they have a low protein requirement (30%). With low density of rearing, they grow well with natural food. There is no expenditure on special feed. It is because of these special features that these two species have been selected and established for culture.

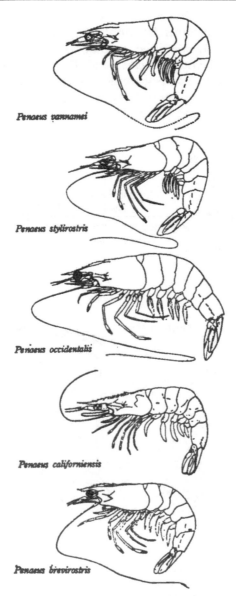

Penaeus vannamei

Penaeus stylirostris

Penaeus occidentalis

Penaeus californiensis

Penaeus brevirostris

Fig. 4.5. Shape of 5 species of the genus *Penaeus* living along the Pacific coast of Central and South America (from Fonseca, 1970).

(2) Culture pond

The ponds during the early period of culture development in the southern part of Ecuador were built by the farmers in the coastal region. Water for these ponds was procured by utilizing the difference between ebb tide and high tide. They thus resemble the milkfish ponds found in Southeast Asia. Once the commercial profitability was recognized, the pond structure was suitably modified so that the yield could be ensured. Figure 4.6 shows the layout of such a pond. The entire operation is mechanized. The water inlet canal is governed by creek conditions. The layout includes a small seedling pond, large rearing pond, inlet and outlet for water, and water-drawing pump. On the whole, the operation is well controlled.

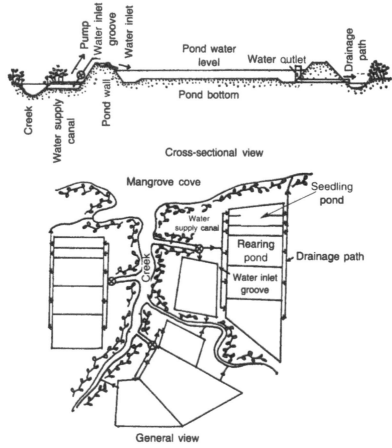

Fig. 4.6 Layout of rearing ponds.

The seedling pond is usually 2-5 ha and the rearing pond 3-60 ha. They are usually rectangular but, depending on ground topography, may be crescent, or semilunar or some other shape.

Culture sites at the southern end of the Gulf of Guayaquil, Peru are shown in Fig. 4.7. It can be seen that they vary in shape.

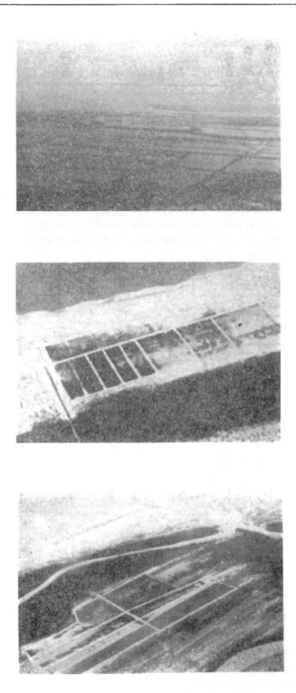

Fig. 4.7. Rearing sites (southern end of the Gulf of Guayaquil).

The pond usually has a peripheral groove or canal for holding the prawns. It is about 5 m wide, 1.5 m deep, with 1.2 m as the water depth in the central part. The structure varies according to ground contour. The water depth may vary from 1 m to 3 m. In certain cases the prawn-holding canal is absent.

(3) Seedlings

As shown in Fig. 4.8, the habitats of mother prawns and the spawning regions are spread along the entire coast. The density of seedlings is high in the shallow coastal mangrove regions and they are readily collected throughout the year. The collection mostly depends on natural seedlings. However, the rapid development of cultural centers in Ecuador is not able to meet the demand in some places due to the higher costs of natural seedlings. This has led to an unstable seedling supply. A few private concerns have set up artificial seedling production centers to supplement the natural seedling supply.

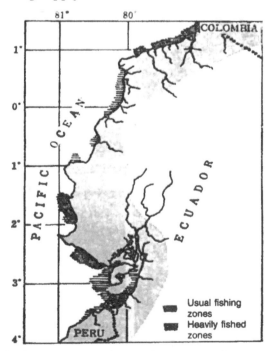

Fig. 4.8. Distribution of prawn fishing sites (SRP, 1977).

In Peru also a private concern has set up a production center. However, since the culture center area is limited, the demand and supply of natural seedlings are still balanced and the production of artificial seedlings has yet to begin.

Natural seedlings are collected by professionals in this field. The seedlings collected by fine mesh nets are transferred to plastic or steel containers and then transported to rearing ponds by boats or small trucks. The size of the collected seedlings varies from about 1 cm body length for postlarvae to 4-5 cm for juvenile prawns. The seedlings are classified as small, medium, and large and sold accordingly. The smaller ones are kept in seedling ponds and the larger transferred directly to rearing ponds.

The density of these seedlings in the ponds depends on their size. As per the standard norms it is 30 seedlings per m^2. The water exchange rate of the seedling pond is higher than that of the rearing pond. Other aspects, such as application of bird droppings or artificial fertilizer and occasional provision of feed such as rice bran or fish meal are the same as in the rearing ponds.

(4) Rearing

When juvenile prawns grown for 1-2 months in the seedling pond attain a body length of 4 cm, they are transferred to the rearing pond. When feeding is not aggressive, the density of juvenile prawns is 3 per m^2.

Management during the rearing period includes water injection by pump for 8-12 hours a day excluding the high tide period, removal of carnivorous animals (perch are netted and waterfowl frightened off by gunshots), partial catching at the time of excessive density for periodic measurements of density and growth, application of fertilizer and supplementary feed when growth is not uniform, and appointing guards to prevent poaching. The period of rearing is 6 months when the target is two rearings a year but, depending on the management and conditions of the pond, it may be 4 months or even 10.

(5) Catching

When the prawns of the pond attain marketable size, preparations are made for the catching operation. Water is gradually drained away and the level in the pond lowered over 2-5 days. Prawns collect in the deepest part, i.e., the peripheral groove (canal) where they are netted. Photos of catching from the rearing pond in Peru are shown in Fig. 4.9.

The prawns removed from the pond are carried to the processing plant in cold-storage vehicles either as such or after beheading in a small room adjacent to the pond. Figure 4.10 shows the beheading procedure and Fig. 4.11 is a photo of the processing plant.

4.2.3 Productivity

The annual production per unit is extremely important in working out the commercial profitability as well as the fish size composition. However, this figure varies widely depending on the site of pond establishment, pond construction, and type of management. Furthermore, ponds

Fig. 4.9. Catching from the rearing pond (Peru).

Fig. 4.10. Beheading.

Fig. 4.11. Processing.

range in production from just 100 kg to more than 1,300 kg. Figure 4.12 shows the wide range of production observed in 21 ponds of Company A in Peru in terms of mean productivity from 1979 to 1981. Although all these ponds are under the control of the same concern, productivity varies widely from a mere 16 kg ha^{-1} $month^{-1}$ (192 kg ha^{-1} y^{-1}) to a high of 55 kg ha^{-1} $month^{-1}$ (660 kg ha^{-1} y^{-1}).

These variations in productivity are due to pond structure and layout as well as type of management. According to a survey conducted by El Rosario Company, ponds covering an area of about 21,000 ha, equivalent to about 60% of the culture ponds in Ecuador, do not have seedling

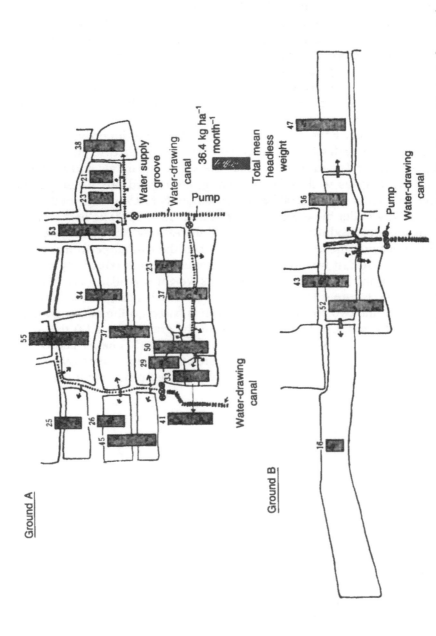

Fig. 4.12. Productivity of various ponds of Company A (Peru). Numbers show mean productivity of headless weight kg ha^{-1} month^{-1}.

preparation ponds. Moreover, the so-called extensive method of culture of 15,000 larvae ha^{-1} is adopted without much water exchange. Hence the annual productivity is 90-450 kg ha^{-1} with a mean of 180 kg ha^{-1}. In Ecuador, 30% of the total production by culture is from ponds that have adopted this method. Ponds in which semi-intensive culture is followed, involving seedling preparation and supplementary feeding, cover about 34% of the total area, that is, 12,000 ha. The productivity of this method is 450-900 kg ha^{-1} with the mean at 550 kg ha^{-1}. Production is about 50% of the total amount.

Several ponds have adopted the modified semi-intensive cultural method with a professionally controlled system. In this case the seedling preparation ponds and rearing ponds are set up separately and effective water exchange as well as forced feeding are carried out. These ponds cover an area of about 2,000 ha, which amounts to 6% of the total area. Productivity under these parameters has been 1,200-1,300 kg ha^{-1} y^{-1} and the quantity of production is expected to increase by 20% of the total.

Now, leaving the above data aside, when the overall performance in 1983 was studied, it was found that the annual mean productivity per ha for Ecuador and Peru in terms of headless weight was 400 kg and 500 kg respectively. Figure 4.13 graphically depicts the changing trends in productivity of the two countries. This graph is based on the data provided in Tables 4.2 and 4.3. It can be seen from the graph (Fig. 4.13) that in 4 years from start-up of prawn culture Peru surpassed Ecuador and occupied a superior position until 1984; thereafter Ecuador, employing the semi-intensive method of culture rapidly increased productivity and surpassed Peru.

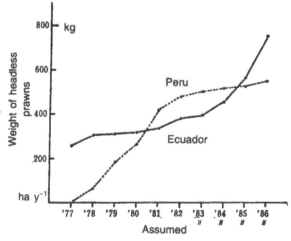

Fig. 4.13. Growth of productivity per ha.

The rapid growth of productivity in the initial period in Peru was due to the semi-intensive culture method followed in all the ponds and excellent management. Depressed growth thereafter was due to apprehension regarding commercial profits given the expenditure on feed required for semi-intensive culture.

The survey made by El Rosario also suggests that the progress in semi-intensive culture made in Ecuador is attributable to raw feed, natural seedlings, and dependence on professional supervisors. Between productivity and profitability, each country has its own characteristics which ought to be carefully considered.

4.3 ANALYSIS OF CULTURE MANAGEMENT IN PERU

4.3.1 Control and Production

The general items of control in the intensive culture method adopted in Peru are as follows: basic items include preparation of the seedlings, their transferal to rearing ponds, acceleration of growth through application of fertilizer, supplementary feed, and final catch. Other items are: pumps for water exchange, machines, tools, maintenance of the pond, prevention of predators, vigilance against poaching, and periodic checks on growth and stock. To implement all these tasks requires a large number of workers. Moreover, sufficient vehicles to transport the workers to their respective places of work, effective placement of large pumps and maintenance of a large quantity of seedlings are mandatory—all of which entails a large capital outlay. Unless the income from production exceeds these expenditures, there will be no commercial profit. Entrepreneurs are working very hard to obtain an effective balance between management and production.

Overall management is directly reflected in duration of rearing and harvest of specified sizes. With reference to the rearing period, Fig. 4.14 shows the conditions of the ponds owned by Company A and the degree of growth of the cultured prawns.

The growth of prawns varies in accordance with the environmental conditions provided and consequently the duration of rearing required to attain a marketable size likewise varies greatly. When the mean body weight is taken as 16 g at the time of catching, the rearing period works out to 5 months and therefore the number of harvests per annum is 2.4. To understand the wide range in size of prawns preferred by traders, the size ratio of prawns reared by 12 companies is depicted in Fig. 4.15.

It can be seen from Fig. 4.15 that the catch of large prawns is rich in the case of Company A because the pond location is very good and management excellent. On the other hand, only small prawns are obtained by Company L due to problems in funding and also lack of

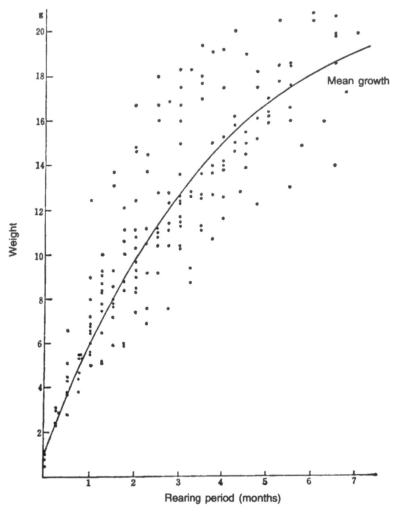

Fig. 4.14. Growth of prawns in ponds of Company A.

experience in culture management. Sizewise analysis of the prawns obtained by all the companies further indicates that the highest ratio of 41/50 is for the smaller prawns. Even smaller ones constitute almost 30%. This reveals that many of the traders are still not serious about proper culture management.

Figure 4.16 gives an idea of the collection activities of these 12 companies and therefore of their catch. It can be seen from the Figure that for most of the companies collection was an ongoing activity throughout the year but catch varied according to number of ponds owned and the pond area. For example, Companies A and B collected almost every month.

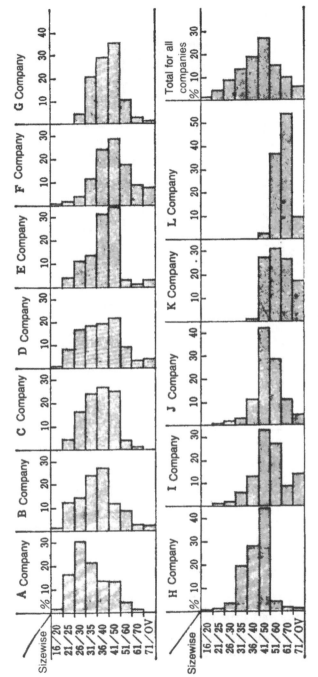

Fig. 4.15. Size composition of prawns preferred by various companies.

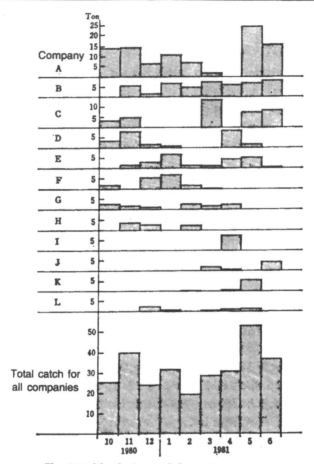

Fig. 4.16. Monthwise catch by various companies.

Companies I, J, and K, on the other hand, collected only for a month or two in a span of 9 months.

Some companies can undertake collection activities regularly throughout the year because natural seedlings are always available and there is not much difference in growth rate between summer and winter. Neither is there much variation in the buy-in price in America, the major export target.

4.3.2 Revenue and Expenditure

The potential production by the semi-intensive culture method was outlined earlier. Now the expenditures involved in production and the extent of commercial profit are discussed.

(1) Expenditure on land procurement

At present the culture ground of Peru is located on the national land adjoining Ecuador. It covers an area of 6,300 ha approved as land for prawn culture development by the Ministry of Fisheries. License for using the land is provided at the rate of 200 ha per applicant. The license fee for 5 years is about 1 yen per m^2 and is payable in advance. After obtaining the permission, if the applicant fails to set up the equipment within 5 years, both license and fees are confiscated. The fee for land usage for setting up equipment (the plant) is 2,000 yen per annum per hectare. Since vast area is required for the culture method, the annual land-usage charges amount to 200,000-300,000 yen. Two or three major traders have acquired a land license for more than 1,000 ha by making multiple applications under different names. If this land has yet to yield, the annual charges of more than 1 million yen together with license fees becomes a heavy burden.

(2) Plant investment

Since the ponds are large, covering a vast area, mechanization is imperative; pond building in 1981 cost about 800,000 yen per ha. Adding the expenditure for water supply canals and grooves for water inlet and drainage, this comes to about 100 million yen to complete a pond of 100 ha. In addition, pump and diesel engine are installed as water-drawing equipment, which with a 20-inch diameter opening and 200 horsepower costs 7 million yen. In fact, for a pond of 100 ha 2 such sets of equipment are needed and thus the total comes to 14 million yen.

In addition to the above, 2 or 3 small trucks (pick-up type costs 2 million yen per truck) are required for transportation. Other expenses include nets, containers, and the tools to operate them as well as for maintenance of a small accommodation for security personnel. Thus the sum required for the complete setting on a scale of 100 ha works out to 120 million yen (1,200,000 yen per ha). The annual redemption fees on all this are also quite high.

(3) Expenditure on production

There are major differences in the scale adopted by entrepreneurs, type of management, and financial arrangements. It is very difficult to determine the standard production costs even for the same land. Therefore details of the expenditure incurred during the first half of 1981 by Company A were analyzed. This company set up 8 pumps for a pond of 300 ha total area and began systematic production. Vehicles include 4 small trucks and 1 medium-size. The total number of workers per month is 50.

The staff composition and their salary structure are as follows: 6 senior staff at the site (graduates in life science, salary 50,000 yen per person per month), 5 drivers (40,000 yen per driver per month), 27 permanent employees (26,000 yen per employee per month). Thus employee salaries total 1,600,000 yen per month. For the month concerned the total sum on production costs, both direct and indirect, worked out to about 6 million yen. The highest ratio went to human expenditure, constituting 31% of the total expenditure. Next comes fuel and gasoline for the engine used in pump operation and for the 5 vehicles, constituting 29% of the total expenditure. Since Peru is a country producing petroleum it gives the impression that fuel charges are quite cheap. However, since the price of heavy oil is 29 yen per liter and that of gasoline 52 yen, expenditure on this account is very high considering the total consumption.

The third major expenditure is the cost of seedlings at about 24%. In the past the seedlings were collected by the entrepreneurs themselves. However, as the scale of the pond increased, the manpower for this collection became inadequate. Moreover, even self-collection became extremely expensive and so some traders specialized in growing and collecting seedlings. Major enterprises began to purchase these seedlings. The traders pack the collected seedlings in containers that are trucked to the culture ponds. The price of seedlings in 1981 was calculated at 30 sen per seedling of large size (3-4 cm), 23 sen per seedling of medium size (2 cm), and 17 sen per seedling of small size (1-1.5 cm). Only about 10% of the seedlings collected are of large size; the majority are small and late postlarval stage. Hence the average seedling price was 20 sen. The number of seedlings purchased is carefully estimated and kept to a minimum. Hence the expenditure on seedlings for a small-scale culture pond is very small. But in the case of a large-scale pond, as in the case of Company A, a massive quantity of seedlings is required (in May, 1981, the maximum amount recorded was 1,500). The expense of seedlings per month ranges from 1 million to 3 million yen.

Thus the expenditure under the three categories, namely, human expenditure, fuel-gasoline expense, and seedling expense covers 84% of the total sum. Repair/maintenance for pump, engine and vehicles constitutes about 8%, the cost of fertilizer 6%, and miscellany 2%.

(4) Revenue income

The average annual production per ha for headless prawns is 500 kg. Thus for an area of 100 ha the yield is 50 tons. The sale price at the site is 1,500 yen per kg and therefore the income on the sale of 50 tons becomes 75 million yen.

(5) Commercial profit

When a pond of 100 ha is established, the ground rent and the investment on the cost of plant establishment work out to about 120 million yen. Next the annual cost of management, worked out as 1/3 that of Company A, is 2 million yen × 12 (months), i.e., 24 million yen. Now the expected income is, as noted above, 75 million yen. The profit therefore is 51 million yen. Calculating the period of investment collection, this works out to 2.4 years. The profit of this business is thus very high.

However, the biggest problem in this case is the percentage of self-funding for various expenses. If the major expenditure is covered by a loan taken from a fund organization, the interest on this loan will be sizable. Hence the period of investment collection will be extended twice or even three times.

4.3.3 Future Problems

Popular opinion in the 1980s held that the demand for prawn in various prawn-shrimp-consumer countries would steadily increase. Since prawn production from the seas would not suffice to meet the demand, production through culture methods attracted more and more attention. Vast low moisture mangrove belts have been/are being developed in the coastal areas of various prawn-producer countries. This will encourage an aggressive production of prawns. It will also improve the life style of the coastal people. However, in Central and South American countries where there is hardly any history of estuarine and marine culture, many important problems in the development of prawn culture must be faced. Unless the following three basic problems are resolved, these countries will never succeed in their efforts to establish effective culture production.

(1) Substantial infrastructure

Freshness is the most important criterion for prawns and shrimps, popular commodities for international export. Therefore it is absolutely essential that the route, transport system, cold storage, freezer vehicles, ice production, and factories for frozen processing involved between time of collection and packing should be of high standard and close to the culturing sites. At present, since these are of low standard, prawns collected from remote places in coastal regions are salt treated and sold in local markets. There is no trace of real culturing.

(2) Public forums to popularize culture methods

Most of the countries have recognized the need for developing culture methods and many have established authorized development centers and educational centers. However, very few countries have taken such fundamental steps for development as an ecological survey of the coasts, practical and technical survey after establishing a pilot farm, and training of technical staff. In many countries neither the people living close to the site nor the capitalists desirous of developing culture production have any idea whatsoever as to where to start or which type of culture method should be adopted.

(3) Fulfillment of financial assistance for development

Many countries have taken practical measures such as tax exemption for those who invest in the culture business and low interest rates by the development fund organization of the government. However, since the fulfillment of items (1) and (2) mentioned above is far from satisfactory and there is no margin even in the government, the net performance is very poor. There are countries such as Costa Rica and Panama where a large amount of foreign money is frequently invested. Ecuador has progressed rapidly due to the involvement of many fisheries and agricultural capitalists as well as technical expertise and Peru enjoys regular funding from capitalists of other industries. Leaving these countries aside, however, regardless of the fact that a country may have a coastal environment suitable for aquaculture, practical development will be difficult to achieve unless the economic conditions are well organized.

REFERENCES

Cerquera OP. *et al*. 1979. *Avances en el estudio de la reproduccion del langostino. INFORME* no. 73, *Inst. del Mar del Peru*.

Corporacion Financiers Internacional. 1979. *Infermacion general sobre aquacultura*. Washington.

FAO. 1977. *Actas del simposio sobre aculcultura en America Latina. Informees de Pesca*, no. 159, vol. 2-3. Rome.

Federation of Overseas Fisheries Associations (FOFA). 1982. Information on prawn fishing in major countries of the world. Kaigyokyo, no. 87.

Fonseca NC. 1970. *Lista de Crustaceos del Peru. INFORME* no. 35. *Instituto del Mar del Peru*.

Hanson JA., Goodwin HL. 1977. Shrimp and Prawn Farming in the Western Hemisphere. Dowden, Hutchinson & Ross, Inc., PA.

Hirono Y. 1982. Preliminary report on shrimp culture activities in Ecuador. Guayaquil, Ecuador.

Holthuis LB. 1980. Shrimps and Prawns of the World. FAO Species Catalogue, vol. 1. Rome.

Minist. de Recursos Natur. y Energeticos. 1977. *Agenda Tecnica*. Ecuador.

Sakamuki N. 1979. Prawn Study and Know-how. Suisansha.

5

Fishing Tools and Methods of Prawn Fishing

Among the various shellfish species, *kuruma* prawns (genus *Penaeus*) and spiny lobsters (genus *Panulirus*) are the most important from the point of view of catch and of production. These two genera comprise many species. *Penaeus* belongs to the swimming type (Natantia) and *Panulirus* to the creeping type (Repantia). As is evident from the terminology, the two genera differ basically in behavior and ecology. Further, the species of the two genera differ considerably in ecological features. Hence the fishing tools and methods also differ depending on the type targeted.

According to the annual statistical report on harvest of prawn species by fishing and culture production (Table 5.1), the fishing catch of prawns in 1981 in Japan was 54,048 tons, which includes 1,064 tons of *Panulirus*, 2,864 tons of *Penaeus*, and 50,123 tons of other species. Little variations from these figures was observed over the next five years (1982-1987). Since the total production of fishing and culture in 1981 (excluding whale catch) was 11,320,000 tons, shellfish constitute only 0.045%. On the other hand, when production is viewed in terms of money expended, the total sum is 821 billion yen, which includes 6,400 million yen for spiny lobsters, 12,800 million yen for *kuruma* prawns, and 62,900 million yen for other species. This sum constitutes 2.26% of the total production cost— 2,777,900 million yen. This shows the great demand for prawn species and their high value.

Table 5.1: Harvest of prawn species between 1977 and 1981 (unit: ton)

Year	Panulirus	Penaeus	Other species	Total
1977	971	2,440	50,016	53,427
1978	1,000	2,673	56,003	59,676
1979	1,064	2,468	49,129	52,661
1980	1,065	2,307	47,133	50,505
1981	1,061	2,864	50,123	54,048

Yet, Japan's import of prawns in 1981 was maximal, i.e., 167,129 tons, costing 279,600 million yen. Even though Japan is renowned for fishery production, still, as far as prawns are concerned, their self-supply percentage is low. This is explained by the fact that areas suitable for fishing in the habitats of prawns in coastal and offshore regions are very restricted and hence there is no scope for widening the scale of prawn fishing.

In other countries, shrimp trawlers which target only prawns are permissible. In Japan, only a mixed catch of fish, including prawns, is allowed or small-scale prawn fishing in the coastal region. Analysis of fishing methods, producing 54,048 tons of prawns in 1981, showed that except for shrimp trawlers there is no large-scale fishing that targets only

Table 5.2: Catch of various prawns in 1981 classified according to fishing methods (unit: ton)

Fishing methods	*Panulirus*	*Penaeus*	Other prawn species	Total
South-seas trawl			4	4
Shrimp trawl			3,525	3,525
Westward bottom trawl			9	9
Westward mechanized boat bottom trawl			1,099	1,099
Midsea bottom trawl			3,756	3,756
Small mechanized boat bottom trawl	5	1,691	30,450	32,146
Broadside sailing boat		3	281	284
Gill net	959	1,060	773	2,792
Stationary net	2	81	117	200
Boat seine		9	4,310	4,319
Others	95	20	5,799	5,914
Total	1,061	2,864	50,123	54,048

prawn species (see Table 5.2).

5.1 SHRIMP TRAWL FISHING

The high-sea bottom seine net used for fishing as carried out by Colombia, Venezuela, Guyana, and Brazil of South America, is popularly known as shrimp trawl fishing. The catch is targeted against prawns for which fishing is done in the midseas off the northern coast of various countries of South America. As shown in Fig. 5.1, as of January 1, 1982 the number of boats registered with the ports of respective countries were: Columbia 14, Venezuela 8, Guyana 19, Surinam 35, French Guiana 22, and Brazil 25 (total: 123 boats).

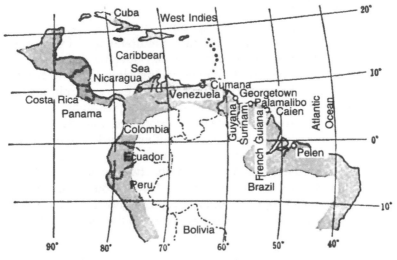

Fig. 5.1. Main regions of prawn trawl fishing in South America.

The fishing boat used is a 99-ton double-rigger trawler (also known as the Floridan trawler). The fishing tools and method are the improved and consolidated type, originally developed in the Gulf of Mexico. These tools and the method are found to be most effective for fishing of *kuruma* prawn (*Penaeus*) and are widely used in prawn fishing worldwide. Fishing operations are carried out by 4 or 5 persons including the captain.

Usually one cruise lasts about 35 days and the catch is not constant for the fishing location. The average annual catch per fishing vessel is 20-40 tons. The main varieties in the catch include pink shrimp (*Penaeus brasiliensis*), brown shrimp (*P. aztecus*), and white shrimp (*P. schmitti*). While these are species of *Penaeus*, they do not include *Penaeus japonicus*, commonly included under the category of other species.

The fishing regions are located midsea (50-100 nautical miles) off the countries along the northern coast of South America. The sea bottom has a gradual slope at a depth of 30-60 m. The amount of catch is given in Table 5.3. Though there was a slight decline after 1976, the catch recovered from 1979.

Table 5.3: Prawn catch (headless and unpeeled) by the shrimp trawl method

Year	Catch (tons)
1976	2,175
1977	1,996
1978	1,997
1979	2,076
1980	2,279

5.1.1 Fishing Tools

The special features of the shrimp trawl method are given in Fig. 5.2. Two sets of otter trawl nets are cast simultaneously by means of an outrigger. At the back of the boat, a small otter trawl net is cast, commonly called the trynet. To determine the density of shrimp habitats and the types, one of the operators raises the net every 20 minutes. For this purpose the fishing tool should be small and lightweight.

The warp (wire rope) is single but its front end attached to a couple of bridles. Each of these is connected to another board. The otterboards are connected to the respective wing nets by pendants on the float side and the sinker side (the pendants extend the two nets). This is the older model fitted directly as an otter trawl fishing tool. The otterboard is attached by the otter rope and quarter rope, as shown in Figs. 5.2 and 5.3. Both nets are used for upward dragging when the fishing net is raised.

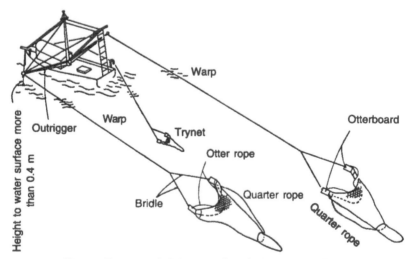

Fig. 5.2. Diagram of shrimp trawling during casting of tow net.

Shrimp trawl nets are broadly classified into three types, as shown in Fig. 5.4. Generally the trawl net is composed of five parts, namely, body net, upper net, lower net, side body net, and bag net. The upper net, also known as the upper part, is composed of a square and batting, the lower net of a beret, the side body net of a side net, and the bag net of a chord head and chord end. The flat net type of shrimp trawl net lacks the dropping ends both in the wing net and the side net. It is connected with a single net having the same mesh up to the front end of the chord head. The balloon net is one in which the mesh of the side net becomes zero at the anterior margin of the chord head or slightly ahead. The semiballoon

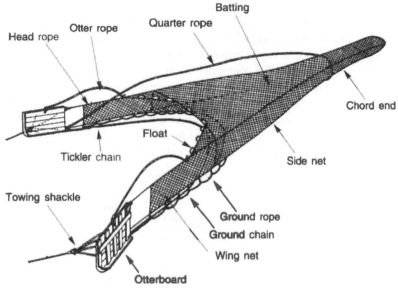

Batting

Quarter rope

Otter rope

Head rope

Chord end

Float

Tickler chain

Side net

Towing shackle

Ground rope

Ground chain

Wing net

Otterboard

Fig. 5.3. Structural diagram of shrimp trawl fishing tools (Tsuya, Collection of Diagrams of Japanese Fishing Boats, 1977).

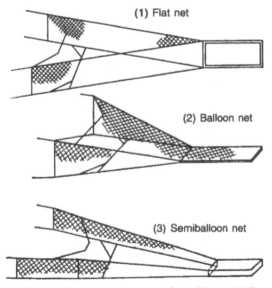

(1) Flat net

(2) Balloon net

(3) Semiballoon net

Fig. 5.4. Types of shrimp trawl net (Nawa, 1967).

net is one in which the mesh of the side net is still left behind up to the anterior margin of the chord head. Hence it is considered intermediate between the flat net and the balloon net.

The commonly used net layout of a 45-ft semiballoon net is sketched in Fig. 5.5. The net layout of a 75-ft flat net capable of increasing the height of the net ends, is sketched in Fig 5.6. The term "foot", expressing the length of the net, actually refers to the length of the head rope (floating net). In Figs. 5.5 and 5.6, percent (%) shows the shrinkage or shortening, m is the length of the head rope or ground rope (floating net), mm the size of the mesh, and 1P, 2B or 1P, 4B expresses 1 point, 2 bars or 1 point, 4 bars. Other Arabic numbers indicate the mesh number. Regarding the term "point" in Japan, the code Kd is used to indicate the net mesh when the two legs are cut and code KS when just one leg is cut (bar).

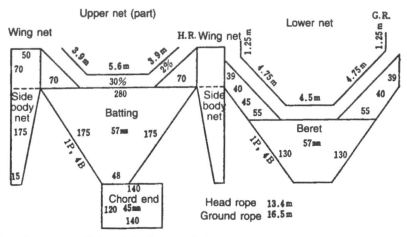

Thread used: polyethylene in body net and wing net—380 D × 24 strands
Chord end: polyethylene—380 D × 57 strands

Fig. 5.5. 45-ft semiballoon net (Kanada, Illustrations of Fishing Tools and Methods in Japan, 1977).

Both the head rope and the ground rope (abbreviated H.R. and G.R.) are compound ropes. The diameter of the head rope is about 12 mm and that of the ground rope 16 mm.

The float attached to the head rope differs according to the structure of the net. Usually a few to ten synthetic floats are used but in the case of a 45-ft net, the total weight of the float is about 10 kg. Compared with other bottom trawls, the buoyancy of the float on the shrimp trawler is small. Since the prawns targeted for catch by the shrimp trawler have a strong tendency to creep and hide in the sand at the sea bottom, there is no need to raise the square of the front margin of the upper net. Therefore a ground chain is attached to the ground rope to play the role of a sinker that stirs the sea bottom and forces the hiding prawns to enter the

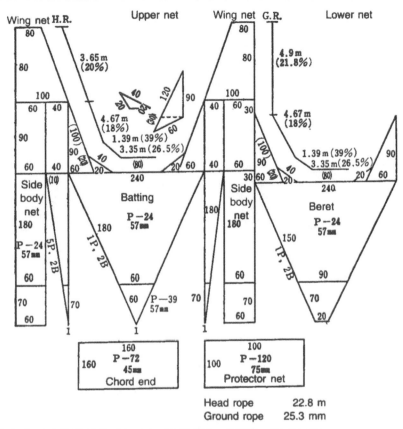

Fig. 5.6. 75-ft flat net (Kanada, Illustrations of Fishing Tools and Methods in Japan, 1977).

net. The method of attachment varies and there are three types, as shown in Fig. 5.7. The size of the link rod of the chain is 6-10 mm (diameter). In (1) of Fig. 5.7, at every 40 cm length of ground chain the link rod is attached at intervals of 30 cm to the ground rope, and at every 90 cm length of ground chain at intervals of 70 cm to the ground rope. In (2) of Fig. 5.7 the chain is fitted to the two ends of the bosom (the front margin of the beret) and also to the center. The chain in this case is larger than that of (1). In (3) of the Figure, the 8-mm chain is more or less of the same length as the ground rope. The ground rope and the ground chain are connected at a gap of about 1.5 m from the suspended chain of 18-20 cm length and 6 mm thick (also known as the foot chain). This method is used in the shrimp trawlers operated in the northern part of the Gulf of Mexico.

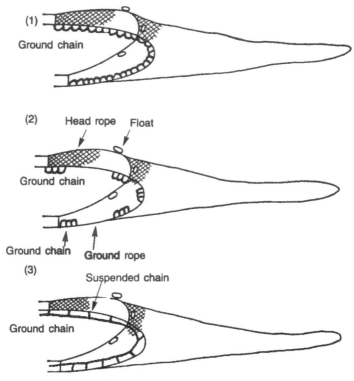

Fig. 5.7. Methods of fitting the ground chain (Nawa, 1967).

Since prawns of the genus *Penaeus* which are targeted for the catch remain hidden in the sea bottom, it is necessary to make them emerge before the nets approach. For this purpose a tickler chain (prawn arousal chain) is used at the front part of the ground chain. The design is shown in Fig. 5.8. As in the case of the ground chain, link rods of 6-10 mm diameter are used in the tickler chain. The chain is 60-90 cm shorter than the ground rope and during the course of net towing is adjusted in such a way that it is 25-30 cm away from the ground chain at the center of the net opening.

As mentioned earlier, the head ropes of the otterboard and wing net are connected to the float pendant and the ground rope to the sinker (see Figs. 5.3 and 5.8). The length of both pendants is about 2 m each. The pendant on the float side stretches the head rope and the pendant on the sinker side stretches the ground rope. The length of the head rope and the ground rope shown in Fig. 5.5 and 5.6 is the length of the net fitted part, and for the actual length of the head rope and the ground rope at the time of setting the net it ظ necessary to extend them by about 4 m each in the pendant part.

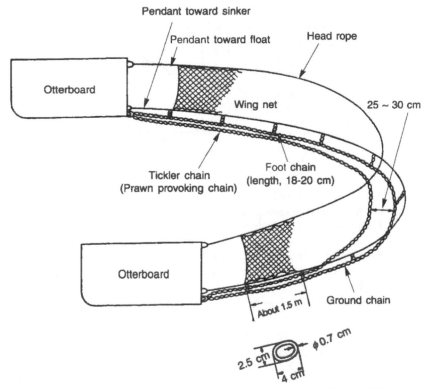

Fig. 5.8. Method of fitting the tickler chain and ground chain (Nawa, 1967).

The otterboard is a net widening board to open the two wing nets horizontally to resist the water current while towing the net. The otterboard used on a shrimp trawler is a flat wooden board. Size varies according to length of the boat. It is 0.9 m × 2.0 m when the boat length is 12-14 m, 1.0 m × 2.4 m for a boat length of 15-17 m, 1.0 m × 2.7 m for a boat length of 17-20 m, and 1.2 m × 3.0 m for a boat length of 21 m. Compared to the otterboard of the wooden boat type, the area has increased 70-90% in steel boats even when they are of the same length. The horsepower has also increased from 200-300 to 380-425. At the same time the scale of the net has widened. The standard otterboard is shown in Fig. 5.9. For bottom otter trawlers other than the shrimp trawler, a vertically long otterboard is used while in the shrimp trawler the old horizontal type otterboard is used. In the case of the old type horizontally elongated board the aspect ratio is 0.47 and in the vertically elongated board 2.5. For better efficiency a board with an inward bent surface is used. The aspect ratio of the otterboard of the shrimp trawler is 0.37-0.45. Efforts are currently underway to reduce this ratio. In the case of a bottom trawl

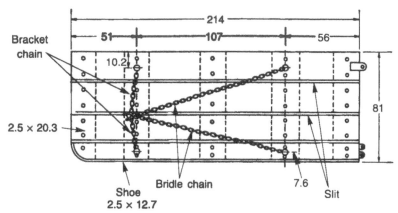

Fig. 5.9. Otterboard used in shrimp trawler (dimension unit, cm) (Kanada, Illustrations of Fishing Tools and Methods in Japan, 1977).

or medium layer trawl the aspect ratio of the otterboard is larger. The otterboard of the shrimp trawler is used for arousing the prawns hiding in the sea floor.

The warp is a wire rope of 12-16 mm diameter. Its front end is connected to a pair of wire ropes called bridles. The bridle is made of the same material as the warp and is 45-50 m in length. Its front end is attached to the bridle chain and the towing shuttle, a bundle of 4 bracket chains (see Figs. 5.2, 5.3, and 5.9).

5.1.2 Fishing Vessels

Strimp trawlers come mainly in three sizes: 70 tons, 99 tons, and 130 tons. The 130-ton model is usually a double-decker while those of lesser tonnage are single-deckers. Regardless of tonnage, fishing tools, fishing methods, fishing equipment, and rigging are practically the same. The fishing equipment and rigging of a 99-ton model (single deck) used in fishing off the northern coast of South America are sketched in Fig. 5.10.

The main mast is centered in the boat. As shown in Fig. 5.10, it may be single form, A-form, or gate form. In all three forms it is used as the supporting pillar for the boom used for lifting the chord end at the time of net raising and outrigging. Therefore it is firmly fixed. Since the warp is connected from the roll winch to the fishing tool through the top roller attached to the front end of the outrigger, the sailing efficiency of the fishing boat during towing is excellent. It takes very little time to prepare for another haul when a good fishing ground of prawns is discovered.

The engine is a Chrysler 425 PS, towing speed 3.0-3.5 knots, and the operations under the control of one man in the steering house. The facilities are such that a single person can handle steering, fish haul, and the telephone.

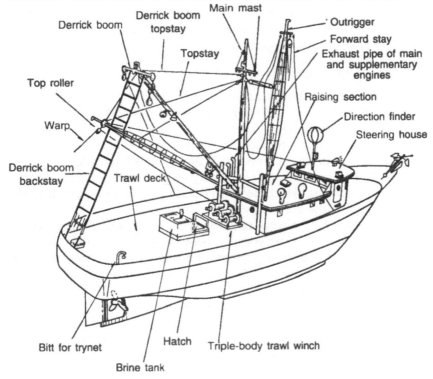

Derrick boom

Derrick boom
topstay

Main mast

Outrigger

Forward stay

Exhaust pipe of main
and supplementary
engines

Topstay

Top roller

Raising section

Direction finder

Steering house

Warp

Derrick boom
backstay

Trawl deck

Bitt for trynet

Hatch

Triple-body trawl winch

Brine tank

Fig. 5.10. 99-ton shrimp trawler.

The trawl winch can be a double-body model or a triple-body (belly) model. In the latter, two of the bodies are used for the right and left warp and the third for the warp of the trynet. For the largest boat (130 tons) a double-body winch for the trynet is fitted separately.

The instruments used for sailing and fishing include a magnetic compass, direction finder, fish finder, remote control device, automatic steering gear, shortwave wireless telephone, weather-related instruments, etc. Compared to fishing boats of about 100 tons operated in Japanese midseas, these devices are not always as effective.

Shrimps and prawns caught by the 70-ton vessel are preserved in ice and carried to land. The 99-ton Japanese trawlers have freezers and the catch is frozen onboard and thus the products remain very fresh. Freezing methods may be broadly classified into two types: (1) a flat tank model or air-cooling model for which the catch is processed into final form aboard the vessel itself; and (2) brine immersion for which shrimps are placed in baskets, immersed in cold brine, then frozen. The first type requires manual labor and is also time consuming. But the product quality is excellent because the shrimps are packed in ice blocks. The second type is simple and the freezing time short; hence it appears to be efficient. However, the products often get damaged and their quality is poor. American fishing boats mostly adopt this method.

5.1.3 Fishing Method

The most important feature of the fishing method by shrimp trawler is the simultaneous towing of 2 other trawls and in addition a small otter trawl called the trynet, i.e., a total of three trawls. The trynet (see Fig. 5.2), being small, can be operated by a single person onboard. The head rope is about 3 m in length. This net is towed ahead and raised at intervals of 20-30 minutes. The catch in the trynet gives an idea of the density of shrimp populations, which helps in deciding whether a particular site warrants dragging the main trawls.

Once the points for main net operations are decided on the basis of the catch in the trynet, the boat speed is slowed and the captain switches steering onto autopilot. Two persons onboard handle casting the nets, another operating the trawl winch, and thus a total of four persons carry out various operations on the trawl deck.

First, the two otter ropes of the otterboard are combined into one. This is passed through the wide block used for lifing the otterboard, which is attached to the derrick boom topstay (see Fig. 5.10) and wound at the trawl winch warp end. As the warp is wound by the trawl winch, the otterboard emerges on the gunwale side. Next the otter rope is reversed, the warp is wound (rolled) and fixed when the length of the 2 bridles becomes approximately 1 m between the top roller and the towing shackle of the otterboard. The other ends of the 2 otter ropes are fastened to their respective head ropes. When these operations relating to the otterboards of the 2 gunwales are completed, the net is thrown out of the gunwale from the chord end. Before this step the loop of the quarter rope is widened and fastened to the chord end. Since the boat progresses slowly, the chord end gets separated from the gunwale and moves to the tail end of the boat. At this stage the batting, beret, and wing are thrown out of the gunwale. This completes the preparations for trawling; details are shown in Fig. 5.11.

At this stage the captain returns to the bridge, turns the boat, and approaches the towing direction at full speed. The brake of the wire drum of the trawl winch is loosened and the warp extended. When the otterboard sinks, the drum brake of the trawl winch is applied and the gap between the otterboards on the two sides of the boat widened to 5-6 m. This is followed by warp stretching (extension). The stretched length of the warp is 3-5 times the water depth. However, during the course of towing the boat turning right has the length of the right side warp stretched to 3-40 cm; the warp drum of the trawl winch is halted by applying the

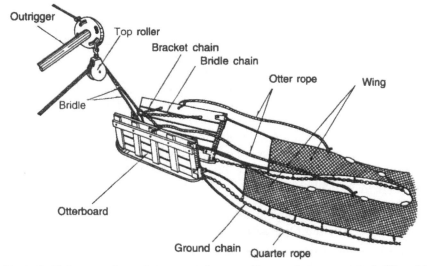

Fig. 5.11. Right gunwale towing preparation or rolling up the warp to the bridle, while lifting the net (Tsyua, Collection of Diagrams of Japanese Fishing Boats, 1977).

brake. When towing is normal, the trynet is cast and the 3 nets towed simultaneously.

The boat speed during towing is 3.0–3.5 knots. As mentioned earlier, the trynet is lifted at intervals of 20–30 minutes to check the quantity of the catch. When the results indicate the existence of concentrated groups of prawns, the boat is turned and towing carried out. While turning, the trynet is lifted. The duration of one round of towing depends on the catching conditions but is normally 4–6 hours. There is always a tendency to prolong towing. Towing is done 2–6 times per day. This is because the prawn catch is poor during daylight hours. Also, to save on fuel and labor, catching operations are mostly carried out at night.

While lifting the net the boat speed is reduced to half and the warp is lifted by trawl winch. Once it attains the condition shown in Fig. 5.11, the boat speed is further reduced and steering put on autopilot. The net lifting operation carried out on the deck is the same as for casting the net. First, the quarter rope is attached to the pole hook and rolled up at the warp end of the trawl winch. When the chord end has been pulled up to the gunwale, the strop is attached to it. By means of fishing tackle the chord end is lifted onto the trawling deck. Since the terminal part of the chord end remains closed by the chord line, the catch is poured out by shaking the chord line. In this state, since the main parts of the net such as the wing, batting, etc. remain outside the gunwale, if the chord line is tied and the chord end cast once again into the sea, the condition shown in Fig. 5.11 is attained. When the boat is moved to change the location of fishing operations, all the fishing implements must be lifted inside the boat. As long as the location is not changed, trawling can be successively continued by the aforesaid method, drawing only the chord end into the boat.

Soon after catching the prawns are beheaded, washed, and frozen. One person can process 20-30 kg per hour. The harvest per day was once about 250 kg. However, the average in Colombia is 108 kg, Guyana 110 kg, Surinam 66 kg, French Guiana 143 kg, Venezuela 161 kg, and Brazil 132 kg. In terms of commercial profit, the base is said to be 127 kg. There is much pressure on management but considering the worldwide increasing demand for prawns and the rapid rise in price, management is finally being well supported.

5.1.4 Prawn Response to Nature of the Net and Fishing Tools Used during Towing Operation

Literature on the fishing tools and the composition of the nets used during towing of the shrimp trawl nets is scant. References are mostly drawn from the results of catch data. The results of one or two model experiments are discussed below.

Details of the report published by Honda (1979), based on the results compiled by Company A, are given in Table 5.4. The shrimp trawl net used in the experiment was the 50-ft (15.24-m) flat net type. The otterboard was 0.8 m × 2.7 m. Against this practically applied net, there is the model net reduced to 1/10. In the model experiment, a vertically circulating water tank was used and the water current in the range of 1.0-3.5 knots (practically converted speed). Under these conditions, the resistance of the net and the gaps between the otterboards were measured. The values given in Table 5.4 are the converted values of the practical case.

Table 5.4: Results of model experiments on 50-ft flat type shrimp trawl net (Honda, Journal of Fisheries Association of Japan, 1979)

Trawling speed (knots)	Gap between otterboards (m)	Resistance of net (kg)	Resistance of otterboard (kg)	Total resistance (kg)
1.0	10.0	174	30	204
1.5	10.5	218	68	286
2.0	11.0	264	118	382
2.5	12.0	406	186	592
3.0	12.0	516	266	782
3.5	12.0	726	368	1,094

Table 5.5 shows the results of model experiments carried out at the Fishing Research Center of the Tokyo University of Fisheries. The shrimp trawl net used in the experiments was a 40-ft (12.2-m) semiballoon net

Table 5.5: Results of model experiments on 40-ft semiballoon type shrimp trawl net

Towing speed (knots)	Gap between otterboards (m)	Central height of net opening (m)	Total resistance (kg)
1.0	7.0	4.2	230
1.5	7.3	3.2	330
2.0	8.0	2.5	470
2.5	8.5	2.2	600
3.0	8.5	1.9	780
3.5	8.5	1.7	1,010

type. The otterboard was a 1/3 reduced model of the actual net (0.60 m × 1.51 m). Since the model net used in the experiment was large, towing experiments were conducted in the sea. The terms of the practical conversion speed lie in the range of 1.0-5.4 knots. Table 5.5 also includes some of the factors pertaining to net composition and various measurements.

Since the net composition differs slightly between the two types mentioned above, it is very difficult to carry out a practical comparison. From the point of view of the gap between otterboards, it appears that the area of the otterboard is small in the case of the 40-ft net. On the other hand, the total resistance of the 40-ft net is slightly higher. These model experiments were carried out in seas with a sand-mud bottom.

In the case of the 50-ft net, experiments were carried out in a water circulating tank. Therefore, differences might have occurred. In any case, in the actual fishing operation, the towing speed is 3.0-3.5 knots. Therefore, in the case of 40- to 50-ft nets, the gap between the otterboards will be 8.5-12.0 m, the total resistance of the fishing tools assumed to be 700-1,000 kg, and the height of the net opening about 2.0 m.

The behavior of shrimps and prawns differs according to the species. The *kuruma* species, targeted by shrimp trawlers, are similar in behavior to *Penaeus japonicus*. According to the observations of Miura on the behavior of *Penaeus japonicus* during the course of towing by a hand-operated net (a type of bottom seine), those which remain inside the sand such as shrimp, flatfish, flounder, etc., are difficult to trap inside the nets even when they approach closely and the sinking net has come into contact with them. This is because the fish are reluctant to emerge. However, when the water is turbid, they do emerge from the sea bottom and get trapped in the net. The shrimp trapped in the net try to escape through it and pass over the head rope.

As already known, shrimps are basically swimming or creeping types. *Kuruma* shrimp 12-24 cm in body length were put in a water tank in which the bottom had been spread with sand. The thickness of the sand was 15 cm. Shrimp behavior was observed when a flat net was moved at

0.50-1.5 m s^{-1} . The results as measured by Takara (1970) indicated that
the maximum speed varied according to body length. The swimming
speed was 0.50-0.54 m s^{-1} when the movement of the flat net was 1.69-
1.77 m s^{-1} . Thus it is evident that the shrimps escape at a speed more or
less close to the towing speed of the shrimp trawler.

A model experiment with 1/8 reduced form of a 40-ft semiballoon
shrimp trawl net (see Table 5.5) was carried out in a rectangular tank in
which the bottom was spread with sand. *Kuruma* shrimp were put in the
tank and their behavior observed vis-a-vis the model net. Since the body
length of the *kuruma* shrimp was only 10 cm (juvenile shrimp) versus the
size of the net reduced by 1/8 of the trawl net, it is not possible to give an
absolute value. However, along with increase in towing speed, the entrance or trapping percentage of the shrimps in the net increased. However, if the otterboard starts floating up and the ground rope separates
slightly from the sea floor due to a high towing speed, the trapping
percentage of the shrimps rapidly decreases. The effect of the tickler
chain is also very significant. In the absence of this chain or when stirring
of the sea bottom by it is negligible, the trapping percentage likewise
decreases rapidly. Therefore, even in the case of actual fishing operations, an increase in towing speed influences the increase in catch rate,
but this rate will decline if countermeasures are not applied to prevent
floating of the fishing tools from the sea floor.

In view of the aforesaid facts, it is clearly imperative to increase
the horsepower of the main engine of the fishing boat or to widen the
area of the otterboard even when the net size is the same, to obtain a
higher towing speed and to increase the sea area for net operation. The
recent trend is to spread the net in a horizontal direction. It can be
concluded that instead of using one 80-ft net for towing, it would be
profitable to use two 40-ft nets. Hypothetically, for a trawl net comprising a 40-ft and 80-ft net, the water current resistance would be propor-
tional to twice the dimension. The rate would become 40^2 and 80^2. The
resistance of the 80-ft net would be 40 times the resistance of the 40-ft net.
However, the width of the net opening during towing of the 80-ft net
would only be twice that of the 40-ft net. Since the height of the net
opening would also double, the area of net opening would increase 4
times. In the case of a shrimp trawler, increasing the height of net opening does not affect the increase in catch rate. Therefore it is desirable to
use two 40-ft nets.

5.2   WESTWARD BOTTOM TRAWL FISHING

Westward bottom trawl fishing is seine fishing in the East Sea and Yellow Sea in a westerly direction at a latitude of E 128° 30' (northward to a
longitude of N 33° 9' 15" it is E 128°) and in the South China Sea, north-

ward at N 10°. The fishing operation consists of two methods, namely, westward trawl fishing and westward mechanized boat seine fishing. Fishing is under strict operational regulations according to the Japan-China Fishing Agreement and the Japan-Korea Fishing Agreement. Problems relating to management have arisen and the number of fishing boats has decreased; from 1978 a catch of only 200,000 tons has been recorded. The permitted number of vessels as of January 1, 1982 included 2 westward trawlers and 439 westward mechanized trawlers. The quantity of catch in 1981 was 487 tons by westward trawl fishing and 183,860 tons by westward mechanized boat seine fishing, for a total of 184,347 tons. The catch included shrimp totaling 9 tons in trawl fishing and 1,099 tons in the mechanized boat seine fishing, for a total of 1,108 tons. The quantity of shrimp catch by westward bottom trawl fishing was 4,510 tons in 1978, 3,349 tons in 1979, and 2,460 tons in 1980. Thus the shrimp catch declined in a 3-year period. In the trawl fishing operation, shrimps are rarely the main target. They are caught mixed with other marine organisms, mainly fishes. Therefore it is advisable to discuss the westward mechanized boat seine fishing operation. In the case of the trawl fishing, one vessel casts one net, i.e., a 1-vessel trawl. In the westward mechanized boat bottom seine, one net is cast by 2 vessels, that is, a 2-vessel trawl. Therefore the number of nets is half the number of trawlers. The manner of towing is shown in Fig. 5.12. There is no otterboard for spreading the net

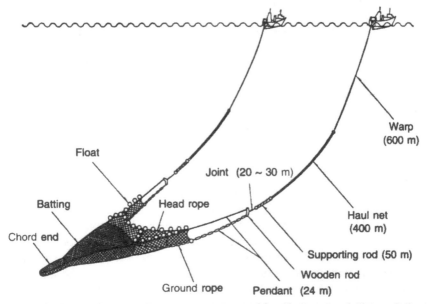

Fig. 5.12. Sketch of towing of westward mechanized boat bottom trawl (Tsuya, Collection of Diagrams of Japanese Fishing Boats, 1977).

horizontally. The net is opened due to the gap between the 2 fishing vessels. Therefore the overall length of the warp, haul net, etc. is far longer than in the case of otter trawl. Related to water depth, it is 1,050 m when water depth is 50-100 m. Figure 5.13 shows the layout of a net targeted to catch shrimps. It can be seen from this Figure that the part between the wing and the side net is in the form of a large triangular net. It is composed of 6 sheets of netting. The structure is considerably different from that of the shrimp trawl net discussed earlier.

The catch made in the westward bottom trawl consists of various types of prawns and shrimps. According to the annual statistical reports on fishing and culture production, the entire catch comes under the category of "others" as it does not include *Palurinus japonicus* and *Penaeus japonicus*. In the westward mechanized boat bottom seine fishing the main type among the shrimp varieties is *Penaeus chinensis*. Though it is a *Penaeus* species, its behavior differs from that of white shrimp, pink shrimp, and brown shrimp, the targets of a shrimp trawler. These shrimps do not migrate much and mostly remain in the sandy sea bottom. But *P. chinensis* makes an annual migration. During the fishing season it is distributed in the middle sea layer and in this aspect differs from *P. japonicus*. Figure 5.14 shows the migratory course of *P. chinensis*. It spawns in all regions along the coast of the Gulf of Chihli and the Yellow Sea; the Gulf of Chihli groups are especially large. In the Gulf of Chihli it spawns in early May and most of the shrimps die in summer after spawning. The shrimps form concentrated groups in the Gulf of Chihli in September-October, then move eastward in November. In December they migrate southward and form a favorable fishing ground in the southern part of the Yellow Sea. In January they descend even further southward to overwinter. In February they once again migrate to the middle part of the Yellow Sea from where they move in large groups northward. In March they are caught in large numbers in the southern waters of Yamato peninsula. In late April they enter the Gulf of Chihli. Therefore the fishing boats have to move with the migrating shrimp groups and towing should be carried out from the bottom layer to the middle in accordance with the vertical migration of the shrimps. This requires an effective fish finder. Regarding the migratory *P. chinensis* there are strict regulations for fishing according to the agreements between Japan and China and between Japan and Korea. The regulations are given in Fig. 5.15. On January 1, 1979, the Agreement on Fishing Operations between Japan and China was partially modified. In order to preserve the fleshy prawn (*Penaeus chinensis*) resources, the prohibited period of operation was extended and again for the preservation of overwintering and juvenile prawns, the protected regions were extended. As a consequence the catch of *Penaeus chinensis* has decreased year after year.

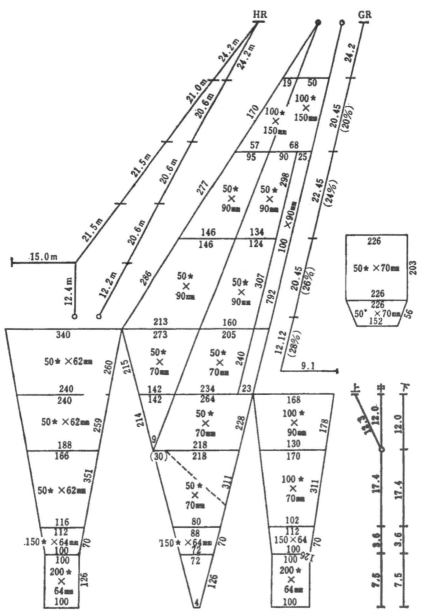

*= strands (units in Japanese for counting the number of ropes)

Fig. 5.13. Shrimp net used for westward mechanized boat bottom trawl fishing.

Fig. 5.14. Migration of *Penaeus chinensis*.

Most of the westward bottom trawl fishing vessels are 100-200 tons. The Japanese vessels are very modern and extremely efficient. To facilitate casting and lifting of the net from the tail-end of the vessel there is a rampway and about 2/3 length of the vessel from the rampway has been made into a fishing deck. The forecastle and bridge lie near the bow.

Of the two fishing vessels, one is called the main vessel and the other, the auxiliary. Besides the captain there are sailors on the main vessel who participate in overall supervision during fishing operations. The

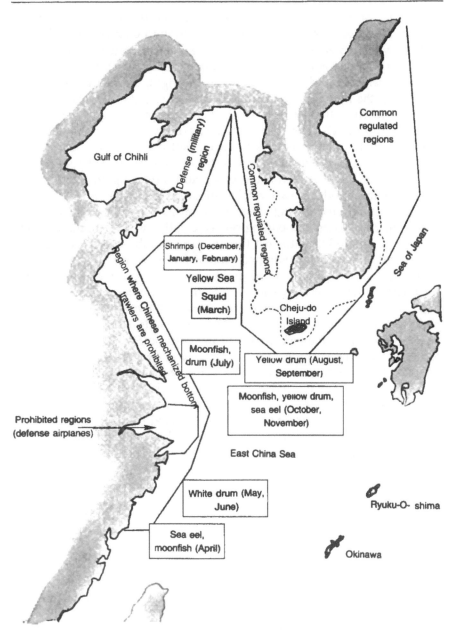

Gulf of Chihli

Defense (military) region

Common regulated regions

Common regulated regions

Region where Chinese mechanized bottom trawlers are prohibited

Shrimps (December, January, February)

Yellow Sea

Squid (March)

Cheju-do Island

Moonfish, drum (July)

Yellow drum (August, September)

Moonfish, yellow drum, sea eel (October, November)

Prohibited regions (defense airplanes)

East China Sea

Sea of Japan

White drum (May, June)

Ryuku-O- shima

Sea eel, moonfish (April)

Okinawa

Fig. 5.15. Main fishing types, fishing period, and fishing sites in the westward bottom seine fishing method.

equipment of the main vessel differs from that of the auxiliary in that fishing and navigational instruments are concentrated more in the former.

Even when towing is done using a single net between two vessels, each vessel has its own net. In the towing direction, when the main vessel is on the left side and the auxiliary on the right, the method of casting and lifting the net used by the main vessel is as shown in Fig. 5.16. The main vessel casts the net from the tail-end rampway while moving forward (Fig. 5.16 (1)). The approaching auxiliary vessel passes over the wooden rod rope and connects its trawl (left gunwale) by means of the rope (Fig. 5.16 (2)). The two vessels move forward at full speed while widening the gap between themselves and stretching the trawl. After stretching the trawl the two vessels maintain a gap of 450-600 m (Fig. 5.16 (3)). Usually the towing speed is about 2.51 knots and the duration of towing 2-2.5 hours. Depending on sea conditions, the duration may be prolonged. Before lifting the net the gap between the vessels is gradually closed and in that state towing continued for 10-20 minutes. At the time of lifting, the two vessels are brought almost alongside each other. The warp end of the auxiliary vessel is passed onto the main vessel (Fig. 5.16 (4)). Using the winch, the main vessel winds up the warp and thus completes lifting of the net (Fig. 5.16 (5)). During this activity the auxiliary vessel prepares for the next round of net casting (Fig. 5.16 (6)). The next round of operations duplicates the first except for taking place on the opposite side. It is thus clear that the warps of the right gunwale drum of the main vessel and the left gunwale drum of the auxiliary vessel are always used and the operation reversed only when towing is carried out by the nets of the respective vessels. This method is known as the single-vessel fishing method. When the warps of both vessels are used, the warps are wound by both vessels, and this is known as the two-vessel fishing method.

5.3   OFFSHORE BOTTOM TRAWL FISHING

Let us now discuss bottom trawl fishing by vessels of less than 125 tons carried out at N 25°, E 128° 30' (when northward to N 33° 9' 15", E is 128°) and in regions westward to E 128°. In Japan, eastward bottom trawling is usually carried out during fishing operations in midseas. Otter trawl is a fishing method imported from Britain, albeit this method of fishing was developed from the ancient Japanese manual midsea net operation method. It has been implemented for about 70 years, from the beginning of the Taisho era (1911-1926). Due to restrictions in the regions for fishing operations, it is not carried out in the Seto Inland Sea nor the East Sea (Tokai). However, it is widely applied throughout the coastal region of Japan and offshore. In 1981, there were 650 single-vessel

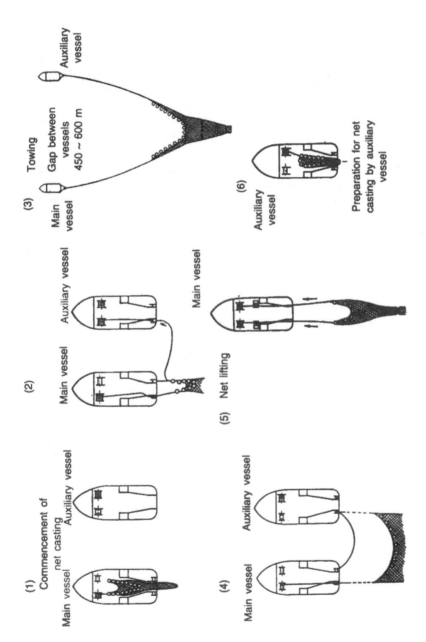

Fig. 5.16. Method of casting and lifting the net during westward mechanized boat bottom seine fishing.

mechanized boat bottom trawlers and 168 double vessels with mechanized boats carrying out bottom trawling (a total of 818 vessels). As in the case of westward mechanized boat bottom trawlers, the double-boat bottom trawling method involves two vessels operating with a single net for towing. In single-vessel mechanized boat bottom trawling, two methods are used—the surrounding trawl fishing method and the otter trawl fishing method. The otter trawl method is allowed only in certain regions, such as the Chiba, Ibaraki, Miyagi, and Iwate prefectures. Therefore in the Hokkaido region, the westward Pacific Ocean region of Aichi prefecture, and in the Sea of Japan, fishing is carried out by the surrounding trawl net method.

In 1981, of the 900,000 tons of caught fish, shrimp constituted only 3,756 tons. This figure included neither spiny lobster nor *kuruma* shrimp. All shrimps were included under the category "others". Midsea bottom seine fishing is not targeted for shrimps but a mixed catch of fishery products.

Two-boat offshore bottom trawling is carried out in regions of the Sea of Japan along Shimane and Yamaguchi prefectures. However, the scale of the fishing vessel is smaller than the westward mechanized boat bottom trawler. The otter trawler is also a small fishing vessel. The surrounding net method mostly uses large nets in the regions around Hokkaido. The offshore regions along Honshu mostly use 50-ton vessels. The fishing method is described in Fig. 5.17. The float is attached to one end of the trawl and the net cast at "A". The boat hauls the net while sailing. At point B the right wing and bagnet are cast and at point C, the left wing is cast. The procedure is then reversed and at point D returns to point A. At this stage towing is carried out by both nets. In about 30-40 minutes of towing, the two nets attain a parallel position. By winding up the seine, one round of operation is completed. In other words, by following A B C D, the "C" bottom is broomed and the fishes forced into the net.

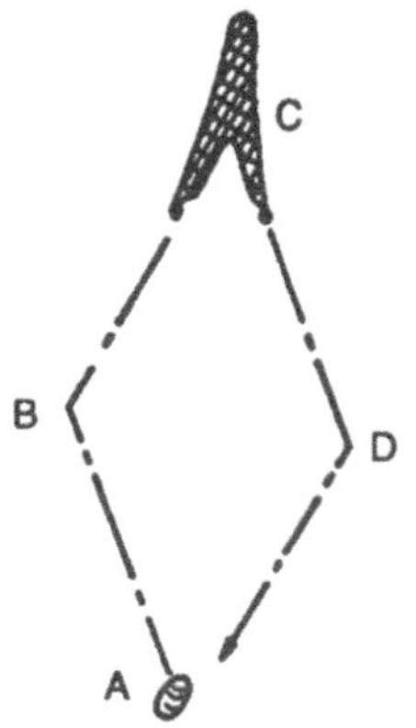

Fig. 5.17. Net casting method in surrounding net bottom trawl fishing with mechanized boat.

5.4 SMALL MECHANIZED BOAT BOTTOM TRAWL FISHING

This type of bottom trawl fishing is carried out by trawlers of less than 15 tons. The operation is classified into different types depending on the fishing tools used. In one type, the fishing tool (net) is towed along the direction of the bow (vertical trawling) and in another type the net is cast along the bow-tail axis and therefore known as the horizontal towing method. Towing carried out by motor boats of less than 5 tons in the Seto Inland Sea is referred to as Seto Inland Sea mechanized boat trawl fishing.

In Kumamoto and Kagoshima prefectures, a total of 169 vessels are carrying out horizontal towing. This method was popular for quite some time throughout Japan. Vertical towing was subsequently preferred throughout the country including the Seto Inland Sea. Of the 28,372 vessels presently engaged in vertical towing, about half, i.e., 12,537 operate in the Seto Inland Sea. Of these, 4,154 weigh 5-20 tons while the remaining 24,000 vessels of less than 5 tons are used for small-scale fishing operations.

The fishing catch recorded for 1981 was 383,738 tons and showed a steady rise for a few years thereafter. The catch included 5 tons of spiny lobsters, 1,674 tons of *kuruma* shrimp, and 24,831 tons of other species of prawns and shrimps in the total catch of 26,510 tons. Compared to other types of bottom trawl fishing, the shrimp catch by vertical towing merits attention.

Let us now look at some of the fishing tools and fishing methods used in bottom trawl collection of shrimps in the Seto Inland Sea. When two nets are towed simultaneously, this is known as the double-net method (Fig. 5.18). The floating net measures 15 m and is spread using bamboo supports. In other words, it is a type of beam trawl. The bamboo used for stretching consists of 2 bamboo poles 4 m in length connected by a vinyl pipe at the center. The two can be widened with a gap of about 7 m. Double towing is carried out as shown in Fig. 5.19. This is identical to the double rigger method of shrimp trawling explained earlier. The trawler is a 4-5-ton boat manned by two persons. The water depth is 20-30 m and towing stretched to 3 times the water depth. In Fig. 5.19 the "rope" is used for manual winding of the net in the boat during net lifting.

Figures 5.20 and 5.21 depict the 4-net type used in towing wherein 2 sets are cast simultaneously and each set is composed of two small nets. The length of the floating net is 5.15 m. Therefore the scale of one set of nets corresponds to about half that of the 2-net type. The 4-net type is shown in Fig. 5.22 but here 3 nets are applied for one towing and the other net used separately. In other words, 2 sets of towing are carried out simultaneously.

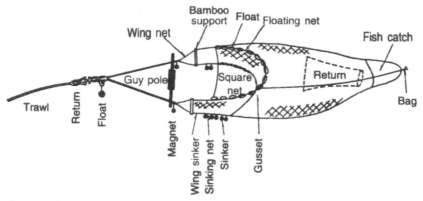

Fig. 5.18. General sketch of shrimp net (double-net type) (Kanada, Illustrations of Japanese Fishing Tools and Methods, 1977).

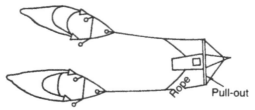

Fig. 5.19. Shrimp net (double-net type) during towing (Kanada, Illustrations of Japanese Fishing Tools and Methods, 1977).

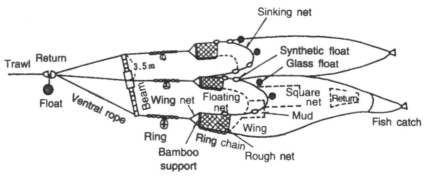

Fig. 5.20. General sketch of shrimp net (4-net type) (Kanada, Illustrations of Japanese Fishing Tools and Methods, 1977).

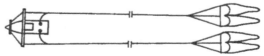

Fig. 5.21. Shrimp net during towing (4-net type) (Kanada, Illustrations of Japanese Fishing Tools and Methods, 1977).

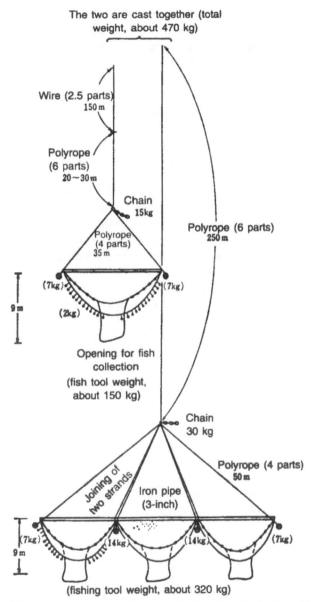

The two are cast together (total weight, about 470 kg)

Wire (2.5 parts) 150 m

Polyrope (6 parts) 20~30 m

Chain 15kg

Polyrope (4 parts) 35 m

Polyrope (6 parts) 250 m

(7kg) (7kg)

(2kg)

9 m

Opening for fish collection

(fish tool weight, about 150 kg)

Chain 30 kg

Polyrope (4 parts) 50 m

Joining of two strands

Iron pipe (3-inch)

(7kg) (14kg) (14kg) (7kg)

9 m

(fishing tool weight, about 320 kg)

Fig. 5.22. General sketch of shrimp net (4-net type) (Kanada, Illustrations of Japanese Fishing Tools and Methods, 1977).

5.5 OTHER SHRIMP FISHING METHODS

More than 60,000 vessels are available for fishing by gill nets. Of these, more than 40,000 are small (less than 3 tons). Gill-net fishing is a small-scale industry along the coast, but the total catch amounts to about 430,000 tons (1981). Therefore it plays an important role in coastal fishing. However, since the total catch of shrimps is 2,792 tons, which includes 959 tons of spiny lobsters, 1,060 tons of *kuruma* shrimp, and 773 tons of others, the percentage of shrimp catch in the total catch is not high. The total catch of spiny lobsters along the Japanese coast is said to be 1,061 tons and that of *kuruma* shrimp 2,864 tons. Therefore gill-net fishing plays an important role in the catch of spiny lobsters and *kuruma* shrimp. Spiny lobsters are caught from the west of Chiba prefecture up to Kyushu along the Pacific coast while *kuruma* shrimp are mainly caught in regions surrounding Seto Inland Sea.

In Wakayama prefecture, the net used for gill-net fishing of spiny lobsters may be a single or triple net. Several nets measuring 26-37 m in length are joined into one set. A small fishing vessel uses several sets. In the case of a single net, there are 6-8 strands of amilan 210 D and the mesh size is 10.2-11.4 cm; there are 13-15 wall nets on flues. In the case of a triple net, the body net has a smaller mesh and the outer nets are composed of 10 strands of amilan 210 D and mesh size 30.3 cm. The gill net is cast before sunset and lifted the next morning.

In Mie prefecture also, *kuruma* shrimp are caught by gill-net fishing. There are two strands of amilan 210 D, 7 sections of 55.5 segments, each measuring 50 m in length. About 30 nets are used. The net is cast between sunset and midnight. Unlike in the case of spiny lobsters, for shrimp fishing the net is lifted 30 minutes to 1 hour after casting and this operation repeated 3 times during the night.

The shrimp basket method of fishing catches 3,100-4,700 tons. Since the fishing tools and equipment are simple in structure, it has received much attention as a fishing method. The shrimps targeted in this technique include *Pandalus borealis* Kroyer, *Pandalus hypsinotus* Brandt, *Pandalopsis japonica* Balss, *Nephrops japonica* Tapparone-Canefri, *Pandalus nippo.·.·nsis* Yokota, *Panulirus japonicus*, and *Penaeus japonicus*. *Pandalus borealis* has recently drawn considerable attention. This species is widely distributed in the North Pacific and North Atlantic oceans and the Arctic Sea. In other words, it is an extreme northern variety. In Japan, it is distributed at a depth of 200-550 m in the Sea of Japan.

Pandalus borealis basket fishing (Fig. 5.23) is done offshore of Niigata prefecture. The basket mesh is 23 mm. The trunk rope and the floating rope are both 16 mm in diameter. One end of the supporting rope of 10 mm is connected to the basket and the other end attached to the trunk rope. The gap of the basket when fixed to the trunk rope is 10 m. About 1,000 m of trunk rope and 1,000 baskets are hung from a chain. Only 5 chains may be used at one time.

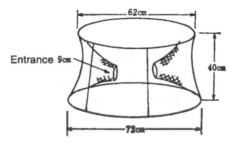

Fig. 5.23. *Pandalus borealis* basket.

The fishing boat is a single-deck 10-tonner. The vessel is taken out early morning and the baskets deposited the previous evening are lifted. The layout of the fishing boat is shown in Fig. 5.24. Item no. 5 in the boat is the location where the roller on the right side of the bow is lifted by the rope from the pickup rope by means of a winding drum. When the baskets are lifted, the supporting rope separates from the trunk rope. Item no. 2 refers to baskets opening and releasing the catch. Item no. 3 shows the preparation for casting the baskets in the next round of fishing. Item no. 4 refers to the transporting of the baskets to the tail end of the boat and item no. 6 the connection between the trunk rope and the supporting rope. By this connection, the pulling and pickup of the baskets along the left side from the tail end of the boat is accomplished. The baskets are cast from the tail end of the boat while it is moving at full speed. Except during bad weather, this is repeated any number of times daily.

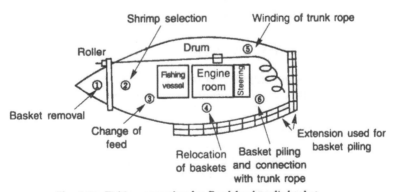

Fig. 5.24. Fishing operation by *Pandalus borealis* basket.

6

Physiology of Prawns

Perhaps no other peoples relish prawns and crabs as much as the Japanese do. It is even asserted that the Japanese people eat prawns and crabs collected from all parts of the world. It is a blessing in disguise for many countries to produce prawns and crabs by fishing or culture with Japan as the target country for export. The percentage of prawns and crabs covering the protein consumption of a Japanese person is at most 2%. Thus instead of claiming that prawns and crabs are an indispensable diet source for the Japanese, they may simply be considered one of the favorite food items. In other words, prawns and crabs are part of the feast as well as a delicacy in Japan. While eggs and potato make one conscious of calories, the attractive appearance of the spiny lobster makes the Japanese housewife completely overlook the caloric factor. Prawns are indeed majestic creatures nutritionally possessing both protein and fat. Though quantitatively less than in other food products, still they are a part of the Japanese nutrition and their value as food cannot be ignored. In this context, it is desirable to know the chemical nature of the various components of prawns and shrimps.

6.1 GENERAL COMPONENTS

Usually the general components refer to water, protein, fat, carbohydrate (sugar and fiber), and ash contents, which are expressed in terms of g per 100 g. Columns 2-7 in Table 6.1 give the details.

The flesh of fresh prawn appears to be full of water. In fact, the water content is 75-80%, which is higher than in chicken meat or fish flesh. The difference of 5% water content between *amaebi* and *ise-ebi*, as seen in Table 6.1, produces a subtle difference in food texture. Obviously the water content decreases during cooking or processing as well as during preservation by freezing.

Compared to chicken meat and fish flesh, prawn meat is poor in fat content. Hence the energy value as a food is also very low.

Table 6.1: Composition of palatable parts of prawns and crabs

Food item	Wa-ste (%)	Ene-rgy (kcal)	Water con-tent	Pro-tein	Carbohydrates			Ash con-tent	Inorganic contents (mg)					Vitamins							
					Fat	Sugar	Fiber		Cal-cium	Phos-pho-rus	Iron	Sodi-um	Potas-sium	A (µg)			IU	B₁ (mg)	B₂	Niacin	C
				(————— g —————)					(————————— mg —————————)					Reti-nol	Caro-tene	A-effi-cacy		(————— mg —————)			
Prawn																					
Amaebi, raw	60	76	80.9	17.0	0.5	φ	0	1.6	55	130	0.2	330	260	φ	0	φ	φ	0.02	0.04	1.0	φ
Spiny lobster, raw	60	104	75.9	21.2	1.5	φ	0	1.4	70	250	1.0	130	380	φ	0	φ	φ	0.01	0.10	1.9	3
Kuruma prawn																					
raw	50	93	77.2	20.5	0.7	φ	0	1.6	50	260	0.8	140	450	φ	0	φ	φ	0.07	0.04	3.3	2
boiled	50	107	73.9	23.7	0.7	φ	0	1.7	55	330	0.4	180	430	φ	0	φ	φ	0.06	0.04	3.8	φ
Spotted shrimp, boiled	0	107	69.0	22.4	1.3	0.1	0	7.2	700	400	1.4	1,700	320	φ	0	φ	φ	0.08	0.05	1.7	0
Shiba prawn, raw	40	66	83.5	13.9	0.8	φ	0	1.8	120	150	2.0	270	240	φ	0	φ	φ	0.01	0.11	2.2	2
Dried prawn, unpeeled	0	295	21.6	63.0	3.0	0.1	0	12.3	2,300	1,200	3.6	1,200	900	φ	0	φ	φ	0.20	0.08	6.7	0
Dried spotted shrimp	0	312	19.5	64.9	4.0	0	0	11.6	2,000	1,200	3.2	1,200	1,200	φ	0	φ	φ	0.17	0.15	5.5	0
Boiled and dried spotted shrimp	0	225	35.2	46.9	2.8	0.1	0	15.0	1,500	860	3.0	3,500	680	φ	0	φ	φ	0.16	0.11	3.5	0
Matted	0	130	58.4	16.0	0.3	15.4	0	9.9	140	190	4.0	3,200	240	φ	0	φ	φ	0.01	0.03	2.0	0
Boiled in soya sauce	0	242	30.9	25.9	2.6	28.0	0	12.6	1,500	400	10.0	2,300	540	φ	0	φ	φ	0.14	0.11	5.0	0
Boiled and canned	0	101	74.9	21.8	1.0	φ	0	2.3	65	210	1.5	540	30	●	0	φ	●	0.01	0.01	2.0	0

(Contd.)

Table 6.1 (*Contd.*)

Food item									per 100 g palatable part											
				Carbohydrates				Inorganic contents					Vitamins							
														A						
	Wa-ste	Ene-rgy	Water con-tent	Pro-tein	Fat	Sugar	Fiber	Ash con-tent	Cal-cium	Phos-pho-rus	Iron	Sodi-um	Potas-sium	Reti-nol	Caro-tene	A-effi-cacy	B₁	B₂	Niacin	C
	%	kcal	(..................................g..................................)						(...................................mg...................................)					(......μg......)		IU	(...................................mg...................................)			
Mysis																				
Raw mysis	0	98	78.0	15.0	3.3	0.9	0	2.8	550	340	4.0	360	320	3	0	10	0.10	0.15	1.8	φ
Dried mysis	0	298	27.8	45.7	10.0	2.7	0	13.8	1,800	1,100	14.0	3,000	1,000	6	0	20	0.10	0.15	5.0	0
Boiled in soya	0	238	29.8	27.5	2.0	26.6	0	14.1	1,400	800	10.0	3,600	460	φ	0	φ	0.12	0.11	2.4	0
Salted	0	51	71.0	10.7	0.5	0.3	0	17.5	50	270	3.0	6,200	210	φ	0	φ	0.06	0.10	1.0	0
Krills																				
Raw	0	94	78.5	15.0	3.2	0.2	0	3.1	360	310	0.8	420	320	180	16	610	0.15	0.26	1.9	2
Boiled	0	86	79.8	13.8	3.0	φ	0	3.4	350	310	0.6	620	200	150	13	510	0.21	0.25	1.4	1
Crabs																				
Blue crab, raw	70	89	78.0	18.9	0.9	0.1	0	2.1	60	170	2.0	320	380	φ	0	φ	0.05	0.03	2.7	φ
Horsehair crab																				
Raw	60	82	78.8	18.8	0.3	φ	0	2.1	75	260	0.5	260	460	φ	0	φ	0.07	0.23	2.3	φ
Boiled	60	84	78.6	19.2	0.3	φ	0	1.9	70	170	0.5	240	200	φ	0	φ	0.06	0.20	2.0	0
Warry crab																				
Raw	70	68	82.8	14.8	0.5	0.1	0	1.8	90	170	0.5	350	290	φ	0	φ	0.24	0.60	8.0	φ
Boiled	70	76	81.2	16.4	0.7	0.1	0	1.6	120	150	0.7	270	200	φ	0	φ	0.04	0.09	1.0	0
King crab, raw	70	81	80.0	15.9	1.3	0.5	0	2.3	130	150	1.2	550	370	φ	0	φ	0.03	0.14	1.0	φ
Boiled and canned																				
Warry crab	0	58	84.8	12.8	0.4	0.1	0	1.9	85	120	0.5	640	24	0	0	0	0	0.03	0.1	0
King crab	0	94	77.0	18.4	1.5	0.6	0	2.5	150	170	1.3	600	150	φ	0	φ	0.02	0.10	0.2	0
Squilla																				
Boiled	5	94	79.5	15.1	3.2	0.1	0	2.1	50	190	1.5	440	280	φ	0	φ	0.30	0.10	2.0	0

Note: φ = minute quantity.
Source: Standard composition chart of Japanese food, 1982. Dept. of Science and Technology.

Protein constitutes about 20% in the raw meat. The exact value varies according to the type of prawn. Details of the proteinic composition are given in the next section.

A fact common with other animals is that the flesh of live prawns contains about 1% carbohydrate called glycogen (the liver contains more than 2%). However, at the time of catching, due to frenetic movement, and again with the passage of time after death, it degenerates. By the time the prawn reaches the consumer there is hardly any glycogen left, only the products of its degeneration.

Fibrous carbohydrate is not present in the muscles. However, the shell contains a special type of fibrous carbohydrate. In the case of spotted shrimp and mysis, the shell is also palatable but this fibrous carbohydrate is not digestible and hence lacks nutritional significance. It is because the value of the fibrous carbohydrate is not included that the tally of the common contents of spotted shrimp and mysis does not total 100.

When a sample of the material is completely burned, ash remains and constitutes the ash content. As shown in columns 8-12 of Table 6.1, the ash is composed of various inorganic substances. Compared to chicken meat, the inorganic matter of prawns has more calcium and sodium and less iron. In the material containing shell, the calcium content is very high. The inorganic content of the body fluid of prawns and crabs is more or less the same as that of sea water.

As for vitamins, prawns are not rich in nutritional value.

There was a time when the value of food was judged by its energy value (calories) and vitamin content only. Nowadays, besides these the digestibility, amino acid composition of the protein, and the composition of the fat are also considered. The food generally consumed is a combination of diverse varieties. The important aspect is its overall nutritional balance. At the same time taste, flavor, color, texture, etc. should also be carefully considered. In this context, prawns and shrimps are a rich important source for adding flavor/taste to the food consumed by the Japanese people.

6.2   PROTEIN

As shown in Fig. 6.1, the musculature of prawns is composed of longitudinal and oblique muscle bundles. The combination is extremely complex. These muscles are made of muscle protein.

Muscle protein contains 20-30% myoprotein, which contains myogen, globulin X, and myoalbumin, which are soluble in water or salt solutions.

Of the muscle protein, 60-70% constitutes myogenic fibers which are soluble in concentrated saline solutions (e.g., 0.6 Mol KCl). They contain myocin, actin, actomyocin, tropomyocin, etc. The fish cake popularly

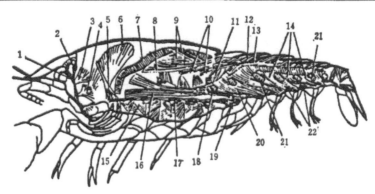

1. Forward moving eye muscle;   2. Backward moving eye muscle;   3. Internal rotating muscle of 2nd antenna;   4. External rotating muscle of 2nd antenna;   5. Anterior dorsoventral muscle;   6. Posterior internal rotating muscle of mandible;   7. Lateral plate drawing muscle;   8. Dorsolateral thoracic-abdominal deep layer muscle (central part); 9. Lateral muscle;   10. Dorsolateral thoracic-abdominal deep layer muscle (lateral part); 11. Dorsolateral superficial muscle;   12. Dorsolateral superficial abdominal muscle; 13. Dorsolateral deep layer muscle;   14. Anterior oblique muscle;   15. Internal rotating muscle of walking leg;   16. Anterior thoracic muscle;   17. Ventrolateral thoracic superficial muscle;   18. Transverse thoracic muscle;   19. Ventrolateral superficial muscle of anterior abdominal appendages;   20.　Anterior oblique muscle;   21. Transverse abdominal muscle;   22. Superficial abdominal muscle

Fig. 6.1. Musculature of *Potamobius* (Zarigani) (Shiino, 1969).

known as *kamaboko* in Japanese contains the myogenic fibers of fish muscles.

In addition to the foregoing, a small quantity of substrate protein which includes insoluble collagen, elastin, and connectin is present.

All proteins are composed of about 18 amino acids. The composition varies slightly in different animals, however. The proteins constituting the human body are the same. Some of these amino acids are termed essential and are dependent upon the food ingested. Table 6.2 shows the standard requirement of the essential amino acids.

The amino acid composition of the muscle protein of prawn compared to fish shows that the essential amino acids—valine, isoleucine, threonine, and lysine—are slightly less and the nonessential amino acids—glutamic acid and aspartic acid—slightly higher (Table 6.3). When these values are compared with those of Table 6.2, it is found that valine alone is slightly lower than the standard value. According to the results of experiments with white rats, the nutritional values such as digestibility of muscle protein, protein utility, and biological value of prawn muscle protein do not differ at all from those of fishes.

The amino acid composition and molecular value of myosin of spiny lobster are the same as the myosin of other animals. The catalytic action which degrades ATP is weak and resembles that of the myosin of scallop.

Table 6.2: Desirable amino acid composition (FAO, 1973)

Amino acid	g per 100 g protein
Isoleucine	4.0
Leucine	7.0
Lysine	5.5
Methionine + Cysteine	3.5
Phenylalanine + Tyrosine	6.0
Threonine	4.0
Tryptophan	1.0
Valine	5.0
Total	36.0

Table 6.3: Amino acid composition of crustacean muscle protein (g per 100 g protein)

Amino acid	*Kuruma* shrimp (*Penaeus japonicus*)	Spiny lobster	Blue crab	Antarctic krill	Fishes
Glycine	4.65	4.63	4.72	4.62	3.0~5.5
Alanine	5.96	5.92	5.72	6.01	5.1~7.3
Valine*	4.39	4.47	4.96	4.72	5.6~9.3
Leucine*	8.62	8.60	8.96	8.28	7.4~9.4
Isoleucine*	3.82	4.08	4.65	5.25	5.0~7.9
Serine	4.18	4.92	4.88	3.91	4.6~5.5
Threonine*	4.10	4.37	5.17	4.20	5.2~6.0
Methionine*	2.83	3.22	2.99	3.44	3.1~3.7
Cysteine*	1.14	1.34	1.65	1.35	—
Aspartic acid	11.70	12.33	11.95	12.5	6.2~11.5
Glutamine	17.51	16.95	16.17	15.3	13.4~16.9
Tyrosine*	4.09	4.11	4.69	4.51	3.5~4.6
Phenylalanine*	4.42	4.72	4.76	6.32	3.4~5.2
Proline	3.66	3.41	4.50	3.35	2.9~4.2
Tryptophan*	1.00	0.95	1.61	1.70	1.1~1.4
Arginine	9.03	7.39	6.29	7.08	5.9~6.9
Lysine*	9.42	9.52	8.89	10.2	9.9~11.8
Histidine	1.92	2.15	2.37	2.40	2.2~3.9
Ammonia	1.30	2.05	1.67		—

*Essential amino acid.

The prawns and shrimps consumed in Japan are mostly frozen products; some are preserved frozen for more than a year. According to the results of studies on spiny lobsters and krills, the myofibril protein of these crustaceans is more susceptible to degeneration compared to the myofibril protein of pollack. As degeneration advances, the tensile strength of all the muscles is lost and they harden. The water-retention capacity declines. The prawn flesh loses its original quality. This type of degeneration occurs even when preserved at –20° C. To prevent the degeneration of myofibril protein of krills, they must be preserved at –40° C and spiny lobster at –30° C. The freezer of the common domestic refrigerator is set at –20° C. Hence it is advisable not to preserve prawns for a long period.

On the one hand, it is believed that the sarcoplasm protein is relatively stable even when preserved. The physical (electrical) features of the various types of proteins constituting the sarcoplasm protein vary slightly in different animals. Such variations are even included sometimes in systematics. Taniguchi carried out electrophoresis of the sarcoplasm protein of prawns in starch gel to identify the type of prawn (Fig. 6.2). This step is requisite in identifying a frozen species since frozen prawns are either headless or completely peeled.

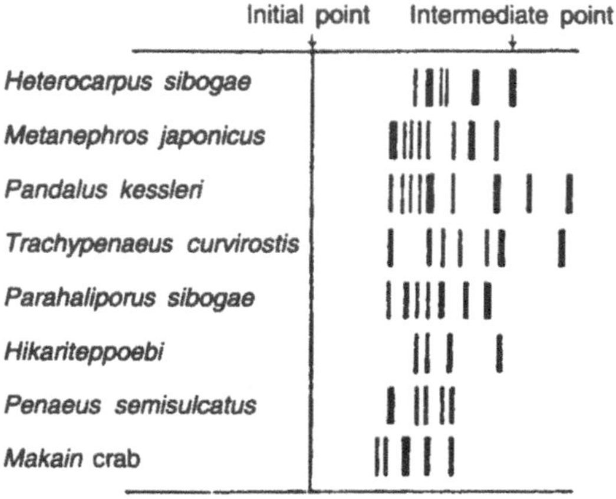

	Initial point	Intermediate point
Heterocarpus sibogae	‖‖ ‖ ‖	
Metanephros japonicus	‖‖‖ ‖‖ ‖	
Pandalus kessleri	‖‖‖ ‖ ‖ ‖ ‖	
Trachypenaeus curvirostis	‖ ‖‖‖ ‖‖ ‖	
Parahaliporus sibogae	‖‖‖‖ ‖‖	
Hikariteppoebi	‖‖ ‖ ‖	
Penaeus semisulcatus	‖ ‖‖ ‖‖	
Makain crab	‖ ‖ ‖ ‖	

Fig. 6.2. Results of electrophoresis of crustacean sarcoplasm protein (Taniguchi, 1979).

The blood of prawns and crabs has a special type of protein, which differs from muscle protein. It is known as hemocyanin. The molecular weight of the hemocyanin of prawns and crabs is about 400,000 or 800,000, which indeed is very large. The molecules contain copper atoms. The binding strength of hemocyanin with oxygen is weaker than that of the hemoglobin of mammalian blood which contains iron. Hemocyanin plays the role of oxygen carrier in prawns and crabs. It has been reported that like sarcoplasm protein, the electrical properties of hemocyanin vary in different organisms. However, it is difficult to apply these variations for differentiating species.

6.3 EXTRACT COMPONENTS

When meat is crushed in water, the various components dissolve in the water. The amino acids, the peptide with 2 or 3 bound amino acids, organic acids, organic bases, nucleotides, sugars, and inorganic substances present in the extract are termed the extract components. In addition to these, water-soluble protein, polysaccharides, certain types of fats, pigments, and vitamins are also extracted. But these are not included in the extract component. The quantity of meat extract varies in different organisms. It is 1-5% in fishes, 7-10% in mollusks, and 10-12% in crustaceans such as prawns and crabs.

6.3.1 Amino Acids

The amino acid composition of the extract of prawns and crabs is given in Table 6.4. In all the species the glycine, arginine, and proline contents are very high in the extract. Each amino acid has its own unique taste (Table 6.5). Their combination assigns a specific flavor to the food. The amino acid composition of prawn extract and crab extract differs in minute detail but otherwise the two have many points in common.

Table 6.4: Amino acid composition of meat extract (mg/100 g meat) (Funita, 1961; Sugiyama *et al.*, 1965; Asakawa *et al.*, 1981)

Amino acids	*Penaeus japonicus*	*Panulirus japonicus*	*Pandalus borealis*	*Euphausia superba*	Warry crab	Blue crab	King crab
Taurine		68	46	206	156	214	264
Aspartic acid	Traces	—	10	52	6	15	3
Threonine	15	6	16	54	4	24	52
Serine	108	107	11	43	3	23	58
Asparagine ⎱ Glutamine ⎰			31	23	10	147	204
Sarcosine			Traces		39	—	—
Proline	230	116	71	217	154	251	226
Glutamic acid	65	7	41	35	11	43	55
Glycine	1250	1078	526	116	253	444	412
Alanine	58	42	90	106	74	144	180
Valine	19	19	23	63	6	48	64
Methionine	11	17	19	34	5	49	43
Isoleucine	11	17	17	48	5	29	54
Leucine	17	12	37	86	8	59	86
Tyrosine	1	11	14	48	5	23	37
Phenylalanine	7	6	18	53	5	20	49
Ornithine			29	42	2	—	5
Lysine	15	21	27	145	8	44	66
Histidine	7	13	6	17	5	23	29
Tryptophan	1		—	—	Traces	6	Traces
Arginine	686	674	181	266	292	329	775
Ammonia			13	18	Traces	4	20

Fujita *et al.* studied the relationship between the amino acid composition of the extract and the taste in several species of prawns. They found that prawns with a delectable flavor have a high glycine content while those with a lower glycine content are less tasty.

Table 6.5: Taste of l-amino acids (Ito *et al.*, 1973)

	Amino acids	Apparent value mg/100 ml	Sweet	Bitter	Pleasant	Sour	Salty
Sweet	Glycine	130	+++				
	Alanine	60	+++				
	Serine	150	+++			+	
	Threonine	260	+++	+		+	
	Proline	300	+++	++			
	Hydroxyproline	50	++				
	Glutamine	250	++		+	+	
Bitter	Valine	40	+	+++			
	Leucine	190		+++			
	Isoleucine	90		+++			
	Methionine	30		+++	+		
	Phenylalanine	90		+++			
	Tryptophan	90		+++			
	Arginine	50		+++			
	Arginine chloride	30	+	+++			
Bitter	Histidine	20	+	+++			
	Lysine chloride	50	++	++	+		
Sour	Histidine chloride	5		+		+++	
	Asparagine	100		+	+	++	
	Aspartic acid	3		+		+++	
	Glutamic acid	5		+	++	+++	
Pleasant	Sodium aspartic acid	100			++		+
	Sodium glutamic acid	50			+++		

Hayashi *et al.* studied the extract of crabs and reported that glycine, alanine, and glutamine assign sweetness while arginine and glutamine together are responsible for the typical "crab" flavor. Proline and taurine, on the other hand, though present in large quantity, hardly contribute to the taste.

Taurine is a sulfur-containing amino acid (Fig. 6.3). Like methionine, also a sulfur-containing acid, administration of a large dose of taurine to an animal effectively lowers the cholesterol level in the blood. Hayashi *et al.* further reported taurine to be effective in lowering blood pressure.

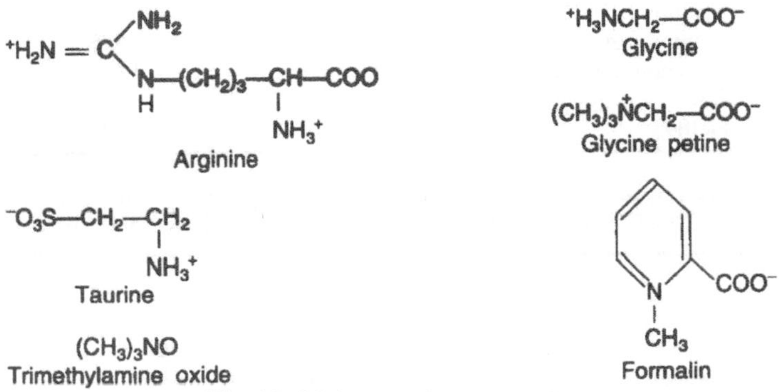

$^+H_2N = C$ $\diagup NH_2$ $\diagdown N-(CH_2)_3-CH-COO$ H NH_3^+

Arginine

$^+H_3NCH_2-COO^-$
Glycine

$(CH_3)_3\overset{+}{N}CH_2-COO^-$
Glycine petine

$^-O_3S-CH_2-CH_2$ NH_3^+
Taurine

COO$^-$
N
CH$_3$
Formalin

$(CH_3)_3NO$
Trimethylamine oxide

Fig. 6.3. Main extract components.

6.3.2 Nitrogen Compounds other than Amino Acids

Glycine petine (Fig. 6.3) is the component responsible for the distinctive "crab" taste of crab extract. This compound is widely distributed in the meat extract of marine organisms. Crustaceans and mollusks contain this compound in large quantity (Table 6.6). In prawns too this compound is assumed to play some role in their distinctive taste.

Trimethylamine oxide (Fig. 6.3) does not contribute directly to taste but is usually present in a considerable amount in marine fishes and mollusks. Prawns and crabs are no exception (Table 6.7). When this compound is reduced by the action of microorganisms it becomes trimethylamine, which gives a strong raw fish smell. The quantity of trimethylamine content along with ammonia content, serves as an indicator of freshness of marine fishes and mollusks. When prawns are treated with sulfite salt, formaldehyde is formed from trimethylamine oxide.

Among the nitrogenous compounds in the extract, the concentration of formalin is high (Fig. 6.3, Table 6.8). This compound is said to be effective in regulating the osmotic pressure of marine invertebrates. It is not detected in freshwater spiny lobster. The relationship between formalin and taste is not clear.

Table 6.6: Glycine petine content of muscles (mg/100 g) (Karasu *et al.*, 1975; Hayashi *et al.*, 1978)

Organism	Glycine petine
Panulirus japonicus	539
Euphausia	365
Warry crab	357
Blue crab	646
King crab	417
Horsehair crab	711
Hanasaki crab	476
Scallop	211
Common oyster	805
Common octopus	1,430
Cod	102
Flatfish	75

Table 6.7: Trimethylamine oxide content of meat (mg/100 g) (Sugiyama *et al.*, 1965; Takagi *et al.*, 1967; Yamada, 1967; Hayashi *et al.*, 1978; Asakawa *et al.*, 1981)

Organism	Trimethylamine oxide	Organism	Trimethylamine oxide
Metapenaeus joyneri	331	*Pandalus borealis*	537
Pandalus nipponensis	363	*Euphausia superba*	212
Trachypenaeus curvirostris	266	*Squilla oratoria*	128
Penaeus japonicus	172	Snow crab	338
Penaeus chinensis	220	Blue crab	140
Panulirus japonicus	213	Horsehair crab	392
Toyama shrimp	318	*Hanasaki* crab	290
Pandalopsis japonica	348	King crab	397

Table 6.8: Formalin content of meat (mg/100 g) (Hirano, 1975)

Organism	Formalin
American crawfish	0
Crangon affinis	104
Pandalus borealis	82
Metapenaeus ensis	105
Metapenaeus joyneri	83
Penaeus chinensis	211
Penaeus japonicus	129
Neomysis intermedia	58
Euphausia sp.	254
Snow crab	167-180
Abalone	140-160
Wreath shell	88-117
Cuttlefish	32-42

6.3.3 Nucleotides

Along with amino acids and their derivatives, nucleotides are important taste-assigning components of the extract (Fig. 6.4). One, adenosine triphosphate, is a carrier of high-energy phosphate in a living organism. After the organism dies, it is enzymatically degraded into adenosine diphosphate → adenosine phosphate (adenelic acid) → inosinic acid (the latter does not form in mollusks). Table 6.9 shows the nucleotide contents of various crustaceans and mollusks. These nucleotides have no taste of their own but strengthen the taste of amino acids and enhance the flavor. They are important constituents of commercial flavor. In fact, the pleasant taste of prawns is attributed to them.

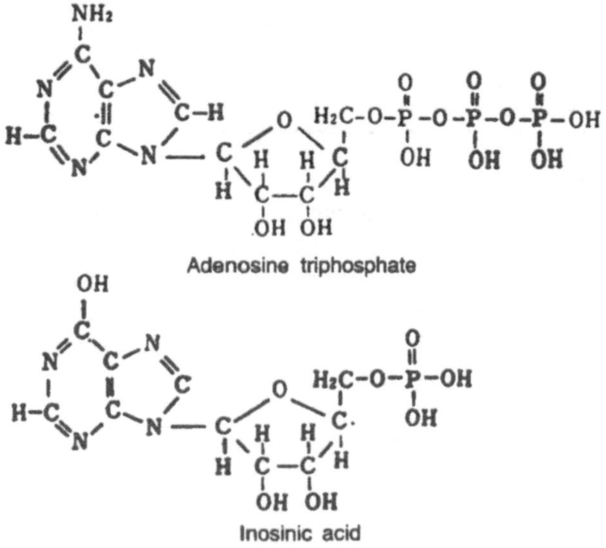

Adenosine triphosphate

Inosinic acid

Fig. 6.4. 5′-nucleotide.

6.3.4 Other Extract Components

As mentioned earlier, in prawns and shrimps the sugar content is either extremely low or almost nil. However, soon after catching, glycogen and the products of its degradations, a variety of sugar compounds—glucose, fructose, and their phosphate esters—appear. In some fishes and mollusks, the high concentration of phosphate ester causes complex changes during processing and preservation. Such cases have not been reported for prawns, however. Generally speaking, the quantity of sugar components in the prawn extract is not of importance. Glycogen, not a component of the extract, is basically a starch. It has no taste but gives a floury texture.

Table 6.9: Adenosine triphosphate of muscles and products of degradation (mg/100 g) (Arai *et al.*, 1961; Hayashi *et al.*, 1978)

Organism	No. of days preserved[1]	Adenosine triphosphate	Adenosine diphosphate	Adenosine phosphate[2]	Inosinic acid
Pink prawn	0	335	89	33	0
	28	9	7	46	56
Horsehair crab	0	298	75	10	0
	35	18	20	120	69
Squilla	0	195	53	33	17
	6	9	12	127	26
Japanese common squid	0	479	65	19	0
	1	42	123	163	0
Scallop	0	176	50	54	0
	1	47	50	53	0
Kuruma prawn (raw)		599	103	12	
(cooked)		2	17	83	92
Snow crab (cooked)			7	32	5

[1]Preserved at –5° C.
[2]Adenilic acid.

Organic acids, such as acetic acid, propionic acid, pyruvic acid, succinic acid, and lactic acid are also extract components. They are present in small amounts (Table 6.10). In the case of little clam (170 mg/100 g) and hard clam (80 mg/100 g), succinic acid is an important taste component. It does not contribute to the taste of *kuruma* prawn (6 mg/100 g) nor of spiny lobster (27 mg/100 g), however.

Table 6.10: Components other than amino acids in the extract of *Pandalus borealis* (mg/ 100 g meat) (Asakawa *et al.*, 1981)

Adenilic acid	8	Ribose		Traces
Inosinic acid	35	Arabinose		Traces
Adenosine	2	Glucose		Traces
Hypoxanthine	14		Na^+	170
Trimethylamine oxide	537		K^+	139
Formalin	304		Mg^{2+}	10
Acetic acid	23		Ca^{2+}	6
Propionic acid	28		Cl^-	241
Butyric acid	5		PO_4^{3-}	358
Malonic acid	6			
Succinic acid	Traces			

Dimethyl sulfide is an odoriferous nitrogen compound and thus differs from the nitrogenous compounds mentioned earlier. This compound is a product of degradation formed by the enzymatic reaction of dime-

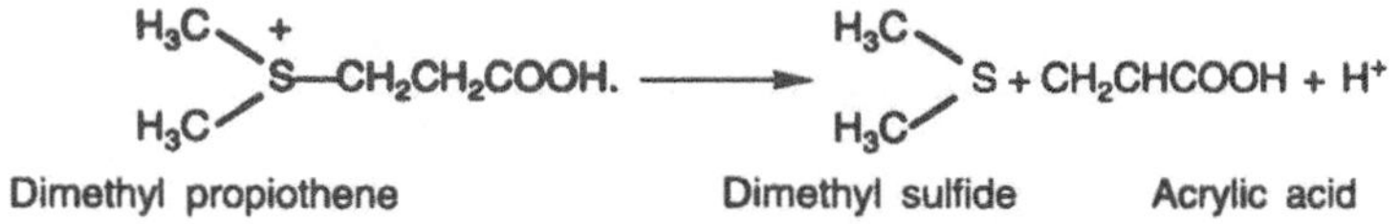

$$\begin{matrix} H_3C \\ \\ H_3C \end{matrix} \overset{+}{S}-CH_2CH_2COOH. \longrightarrow \begin{matrix} H_3C \\ \\ H_3C \end{matrix} S + CH_2CHCOOH + H^+$$

Dimethyl propiothene   Dimethyl sulfide   Acrylic acid

Fig. 6.5. Formation of dimethyl sulfide.

thyl propiothene (Fig. 6.5). Its odor is pungent and human beings can sense it at a minimum concentration of 0.33 ppb (0.33 μg/100 g). The quantity of this compound present in prawns is shown in Table 6.11. Obviously, it is one of the odoriferous components of prawns. During decomposition hydrogen sulfide (H_2S) is produced.

Table 6.11: Dimethyl sulfide and dimethyl propiothene contents (μg/100 g) (Tokunaga et al., 1977)

Organism		Dimethyl sulfide	Dimethyl propiothene
Pandalus borealis	muscles	1.6	97
(preserved in ice)	head	9.1	71
Pasiphaxa japonica	muscles	2.3	540
(preserved in ice)	head	14	940
Argis tar	muscles	1.3	1,500
(preserved in ice)	head	9.9	1,600
Penaeus chinensis	muscles	9.8	30
(frozen)	head	32	41
Euphausia superba	muscles	5.3-9.3	700-37,000
(fresh frozen)	head	9.6-370	2,000-31,000
Euphausia superba	muscles	1.5-3.2	510-27,000
(boiled and frozen)	head	1.1-4.6	670-21,000
Palaemon paucidens	muscles	7.1	165
	head	13	142
Macrobrachium	muscles	6.5	270
nipponense	head	6.6	175

Regarding inorganic substances, as mentioned earlier, the ions of sodium, potassium, chlorine, and phosphate cannot be ignored in respect of extract components of prawns and crabs (Table 6.10).

Summarizing all the points mentioned thus far, Hayashi et al. prepared an artificial extract similar to the natural crab meat extract (Table 6.12). An artificial extract of prawn following the same procedure could also be prepared.

Table 6.12: Composition of synthetic snow crab extract (mg/100 g) (Hayashi, 1979)

Taurine	243	Cytidylic	6
Aspartic acid	10	Adenilic acid	32
Threonine	14	Guanilic acid	4
Serine	14	Inosinic acid	5
Tyrosine	77	Adenosine triphosphate	7
Proline	327	Adenosine	1
Glutamic acid	19	Adenine	26
Glycine	623	Hypoxanthine	7
Alanine	187	Inosine	13
Aminobutyric acid	2	Guanine	1
Valine	30	Cytidine	1
Methionine	19	Glycine petine	357
Isoleucine	29	Trimethylamine oxide	338
Leucine	30	Formalin	63
Tyrosine	19	Glucose	17
Phenylalanine	17	Ribose	4
Ornithine	1	Lactic acid	100
Lysine	25	Succinic acid	9
Histidine	8	Sodium chloride	259
3-methyl histidine	3	Potassium chloride	376
Tryptophan	10	Sodium phosphate (1)	83
Arginine	579	Sodium phosphate (2)	226

6.4   FATS

6.4.1   Composition of Fat

As already mentioned, the fat content of prawns is almost negligible compared to fish meat and cattle meat. The value varies in different parts of the body (Table 6.13). For example, the fat content of the hepatopancreas is 7 times that of the muscles.

Table 6.13: Fat and cholesterol contents of *kuruma* prawn (*Penaeus japonicus*) (Kanasawa, 1976)

Body parts	Fat (%)	Cholesterol (mg/100 g)
Whole body	1.10-1.21	192-220
Muscles	0.77-0.78	120-160
Hepatopancreas	2.37-5.09	340-560
Eyes	4.39-4.57	1,110-1,700
Integument	1.49-1.73	350-440
Exoskeleton	0.34-0.59	60-150
Alimentary canal	1.86-2.30	310-390
Gills	0.97-1.95	210-280
Heart		300-400

The fat is classified into several chemical compound groups (Table 6.14). The ratio of the components varies according to the type of organism and organs and tissues.

Table 6.14: Fat composition of prawn (%) (Kayama, 1982)

Type of fat	*Sergia lucens*	Deep-sea prawn
Wax ester	17.9	60.4
Diacylglyceryl ether	5.8	10.2
Triglyceride	12.6	14.4
Free fatty acid	9.6	4.2
Fatty alcohol	4.4	3.5
Sterol	22.2	3.9
Partial glyceride	8.2	1.8
Phosphatide	19.3	1.4
Fat content	1.7%	13.2%

The wax ester is a combination of alcohol rich in carbons and fatty acid and normally is not present in a large concentration. However, in deep-sea prawns the fat content is high and plays a role in providing buoyancy.

Diacylglyceryl ether is an esterized bond of glycerine with two fatty acids and an ester bond of one alkaline base. The exact function is not clear.

Triglyceride is an ester bond between glycerine and three fatty acids and the oil generally used in food belongs to this group.

The free fatty acids are those not bound with other substances.

The fatty alcohol is the reduced form, $-CH_2OH$ of COOH of fatty acid.

In the case of prawns most of the sterols are cholesterol.

Partial glyceride is derived from triglyceride from which one or two fatty acids are removed.

Phosphatide is a compound in which one fatty acid of triglyceride is replaced by a phosphate compound.

6.4.2 Fatty Acids

As explained, the fatty acids combine with other compounds and form various fats. The fat in a food is an important energy source next to carbohydrates. In terms of wholesome food, it is known that as a source of fat supply fishes and mollusks are wanting. But from the point of view of varieties of fatty acids constituting the fat, fishery products including fishes, mollusks, and prawns are nutritionally important. Table 6.15 shows the composition of fatty acids in the fat of *kuruma* prawn. On the whole there is a similar pattern in the fatty acid composition of fishes and mollusks and they are rich in high-grade unsaturated fatty acids (cattle meat is rich in saturated fatty acids).

Linoleic acid, linolenic acid, and arachdonic acid (Fig. 6.6) are essential fatty acids for human beings. The first two are rich in vegetable oils. In our body arachdonic acid forms from linoleic acid while eicosapentanoic acid and docosahexanoic acid form from linolenic acid. Thus if the first

Table 6.15: Fatty acid composition of *kuruma* prawn fat (Guary *et al.*, 1974)

No. of carbons	No. of unsaturated bonds	Composition ratio (%)	No. of carbons	No. of unsaturated bonds	Composition ratio (%)
Less than 15		6.2	20	2	1.2
16	0	15.4		3	1.3
	1	6.9		4[3]	5.1
	2	1.7		5[4]	13.1
17	0	2.0	22	3	0.4
18	0	6.5		4	2.2
	1	9.0		5	3.0
	2[1]	2.0		6[5]	7.6
	3[2]	0.4	24	2	0.8
	4	2.0		4	4.0
19	0	0.3		5	0.3
20	1	7.9			

[1]Lenoleic acid.
[2]Linolenic acid.
[3]Arachdonic acid.
[4]Eicosapentanoic acid.
[5]Docosahexanoic acid.

$$CH_3(CH_2)_4CH = CHCH_2CH = CHCH_2CH = CHCH_2CH = CH(CH_2)_3COOH$$
Eicosatetranoic acid (arachdonic acid)

$$CH_3CH_2(CH = CHCH_2)_5(CH_2)_2COOH$$
Eicosapentanoic acid

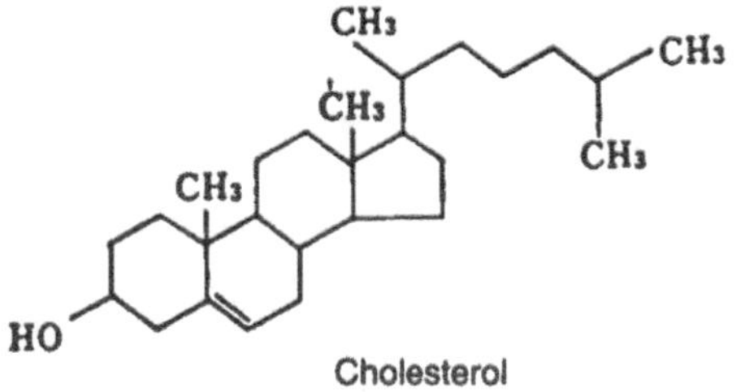

Cholesterol

Fig. 6.6. High-grade unsaturated fatty acid and cholesterol.

Note: Since 1UPAC has changed icosa to eicosa, it is so spelled here.

two essential fatty acids are taken through food there may not be any need to take the other two from an external source. However, the quantity converted within the body is not always sufficient. Arachdonic acid and eicosapentanoic acid, which are high-grade unsaturated fatty acids, can get transformed into a group of compounds which have a physiological activity similar to that termed prostaglandine in the body. The group has compounds which resist agglutination of blood platelets and bring down the cholesterol concentration. It has been confirmed that large doses of high-grade unsaturated fatty acids reduce the cholesterol concentration.

It is clear from the above that though quantitatively the fat of prawns is low, the fatty acid composition is similar to that of fish oil and it possesses excellent physiological qualities.

6.4.3 Cholesterol

It is now confirmed that the cholesterol considered to cause illness in adult humans is nevertheless essential as a source material for the synthesis of bile, which is indispensable for the digestion of fats. In the case of a healthy body, cholesterol is synthesized in sufficient quantity and a constant level is maintained. Moreover, even when cholesterol is taken in large quantity, its level in the body is not disturbed. However, in certain types of disease and also with aging, cholesterol increases in the blood or accumulates in the blood vessels. Since saturated fatty acids increase the cholesterol concentration, it is desirable to control the intake of cattle meat. Again, it is also necessary to control the intake of cholesterol. The cholesterol content of prawn flesh cannot be deemed as low but compared to eggs and liver it is definitely less (Table 6.16). As already mentioned, the fat of prawns contains high-grade unsaturated fatty acids which reduce the cholesterol level and the extract has an amino acid (taurine) with the same action. No harmful effect is produced when prawns are consumed along with fishes rich in high-grade unsaturated fatty acids.

Table 6.16: Cholesterol contents of food items (mg/100 g) (Hisamura *et al.*, 1980)

Organism	Cholesterol
Penaeus japonicus	164
Penaeus chinensis	132
Pandalus borealis	53
Snow crab	72
Japanese squid	180
Oyster	76
Eggs (yolk)	1,030
Cod roe	446
Tuna	50
Chicken liver	391

6.5 SHELL AND PIGMENT

6.5.1 Structure of Shell

The entire body of prawns and crabs is enveloped in a hard shell, also known as an exoskeleton. It lends support to the body and in this context is similar to the skeleton of the human body (Fig. 6.7).

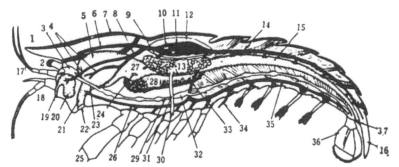

1. Rostrum; 2. Compound eye; 3. Brain; 4. Branch of antennal artery; 5. Stomach; 6. Artery to eye (frontal artery); 7. Shell; 8. Antennal artery; 9. Hepatic artery; 10. Ostium; 11. Heart; 12. Pericardium; 13. Ovary; 14. Posterior artery (supraintestinal artery); 15. Intestine; 16. Telson; 17. 1st antenna; 18. 2nd antenna; 19. Opening of antennal gland; 20. Antennal gland; 21. Esophagus; 22. Mouth; 23. Circumesophageal connective; 24. Opening of left hepatic duct into stomach; 25. Chelate legs; 26. Ventral artery; 27. Posterior stomach; 28. Liver; 29. Ventral nerve; 30. Gonoduct; 31. Oviduct; 32. Thoracic nerve ganglia; 33. Descending artery; 34. Pleopod I; 35. Ventral abdominal artery; 36. Uropod; 37. Anus

Fig. 6.7. Lateral view of inner part of female *Astacus* (Shiino, 1969).

The structure of the exoskeleton of prawns and crabs is shown in Fig. 6.8.

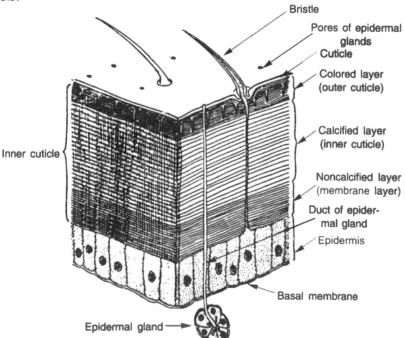

Fig. 6.8. Model diagram of epidermis of prawn and crab (Dennel, 1960).

Next to the fibrous basal membrane is the epidermis and external to the epidermis is the exoskeleton. The innermost layer or noncalcified layer is made of chitin and protein. On its outer side is the inner cuticle or calcified layer, which is extremely thick in crabs. Calcium carbonate crystals are uniformly scattered between the chitin and the fibers made of protein and mucopolysaccharides.

The external side is the outer cuticle or colored layer. It consists of proteins, chitin, mucopolysaccharides, and fats. The proteins in this layer, as explained later, are hardened by quinone. The external coloration of spiny lobsters and others is due to the pigments present in this layer.

The outermost covering is the surface cuticle, comprised chiefly of protein and fat. It is hardened in the same way as the outer cuticle.

Compared to crabs and spiny lobsters, the *kuruma* prawns have more inorganic crystals in the exoskeleton. Moreover, the exoskeleton is relatively transparent and the color of the pigment cells of the epidermis clearly visible.

6.5.2 Composition of Shell

The composition ratio of the components constituting the shell varies in different layers. The overall condition is given in Table 6.17.

Table 6.17: Components of shell (g/100 g dry matter free from fat) (Yanase, 1975; Yano, 1982)

Organism	Chitin	Protein	Ca	$CaCO_3$	Mg
Metapenaeus joyneri	32.4	29.4	15.3		0.6
Euphausia superba	38.7	28.0	13.3		1.0
Crayfish			25.6		
Horsehair crab	18.4	10.5	21.3	51.7	1.2
Blue crab	9.0	6.5	25.3	62.4	2.1
King crab	10.4	44.7	8.2	24.9	

Like prawns and crabs, insects too possess chitin rich in a certain type of polysaccharide. Chitin is a high molecular compound with a linear chain of β-1, 4-N-acetylglucosamine. The larger chains comprise about 10,000 segments of N-acetylglucosamine (Fig. 6.9). As an unutilized source, it has underlying content similar to cellulose. This has encouraged the synthesis of various derivations with chitin as the source material. Intensive studies are underway to work out the possibilities of utilizing the products for industrial and medicinal purposes.

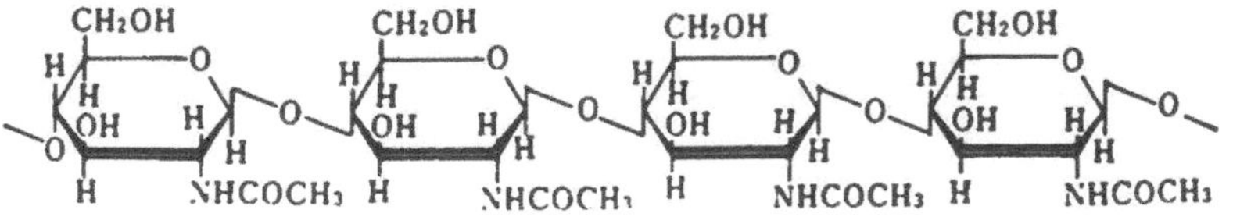

Fig. 6.9. Structure of chitin.

When chitin (Fig. 6.10 (1)) is heated with caustic soda, it converts into chitosan (2). When this is dissolved in acetic acid and made into a film, chitosan acetate (3) film is obtained. When this is treated with mild alkali, it becomes chitosan (2) film. Further, if it is acetylized or acylized, it becomes acetyl chitosan, that is, chitin (1) film or acylchitosan (4) film. Further analytical progress is expected in the near future.

As mentioned earlier, the protein of the shell is hardened by quinone. First of all, a catechol-like substance (phenol with two —OH) and oxidizing enzyme are carried by the blood and an oxidation reaction occurs. The product of the reaction, that is, quinone (—OH becomes =O), bridges —NH_2 of the protein.

The other important component of the shell is calcium. It is evident from Table 6.17 that most of the calcium is in the form of calcium carbonate.

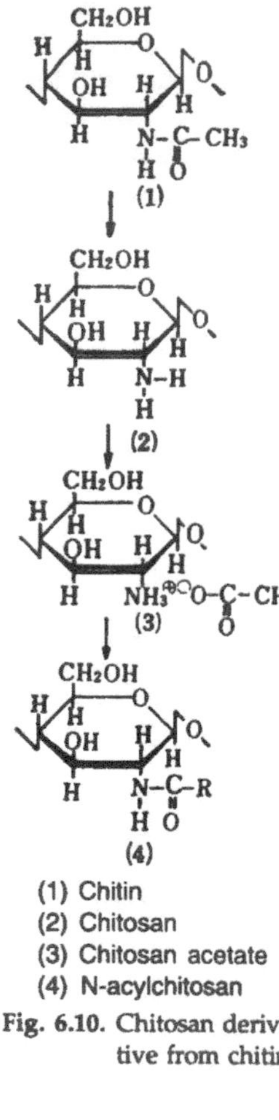

6.5.3　Pigment

The color of prawns and crabs is chiefly attributable to the presence of the caratenoid called astaxanthin. Depending on the species of prawns and crabs, other pigments such as cantaxanthin, phenicoxanthin, and echinonen, etc. are also found (Table 6.18). In the shell there is a small quantity of pigment bound with the protein (carotenoprotein).

(1) Chitin
(2) Chitosan
(3) Chitosan acetate
(4) N-acylchitosan

Fig. 6.10. Chitosan derivative from chitin.

Table 6.18: Carotenoid composition of the shell (%) (Katayama, 1978)

Carotenoid	*Penaeus japonicus*	*Panulirus japonicus*	Blue crab	*Ibacus ciliatus*	*Squilla*	Taka shrimp	*Penaeus chinensis*
β-carotenes	Traces	—	Traces	—	—	—	—
Echinonen	Traces	—	Traces	6	1	—	—
Cantaxanthin	Traces	12	14	2	1	—	—
Lutein	Traces	—	—	—	—	—	—
Zeaxanthin	Traces	—	—	—	—	—	—
Phenicoxanthin	5	2	2	1	0.4	—	—
Astaxanthin	90	48	84	90	95	99	97

Carotenoprotein is blue or purple but when heated turns red. In the past this was explained as due to the snapping of the bond with protein, whereby the free astaxanthin becomes transformed to astacin. However, later observations indicated that while the pigment could be extracted by heating, this did not indicate the formation of astacin. On the contrary, it was found that astacin itself gets decomposed by heating, thereby giving rise to doubts relating to the earlier explanation.

Sometimes blue-colored prawns are found. This is because the total carotenoid content is low and the carotenoprotein concentration high.

The eggs of prawns and crabs have a fresh flesh color. This is due to the presence of astaxanthin. Depending on the species there are other types of carotenoids (Table 6.19).

Table 6.19: Carotenoid composition of ovaries (%) (Miki et al., 1982)

Carotenoid	*Penaeus chinensis*	*Pandalus borealis*	Snow crab	Blue crab
Astaxanthin	75-84	74-76	89-90	95-97
Doradexanthin	5-9	24-26	6	1-2
Zeaxanthin			1	
Idoxanthin	2-5			
(Total content, mg/100 g)	(2.19)	(1.30)	(5.91)	(2.36)

6.6 BIOCHEMISTRY

6.6.1 Nutritional Requirement

Kuruma prawn (*Penaeus japonicus*) culture is being carried out in several regions. However, literature on the biochemistry of prawn has yet to be consolidated.

Regarding the fat requirement of *kuruma* prawn, it is clear that high-grade unsaturated acids, linoleic acid, and linolenic acid are essential. Since prawns lack the ability to synthesize cholesterol, it is also an essential component of the diet and, obviously, so is phospholipid.

Regarding the amino acid composition of protein, it is desirable to take the composition of protein of clams as the standard. It has also been noted that glycine and taurine accelerate feeding.

The effect of sugar in the feed has yet to be determined.

The body color of prawns is an important factor influencing their commercial value. It has been observed that astaxanthin is synthesized in the body according to the steps shown in Fig. 6.11. It is reported that when astaxanthin or zeaxanthin is added to the diet of goldfishes or carps, the body becomes flesh-colored.

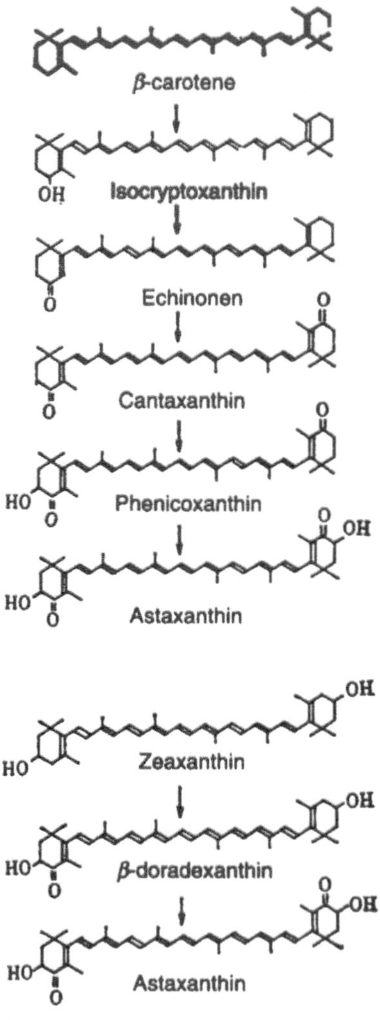

Fig. 6.11. Astaxanthin synthesis in the prawn body (Tanaka et al., 1976).

6.6.2 Enzymes

As only a few reports on the enzymes of prawns and crabs are available, just two or three cases are discussed here.

At the processing factory of prawns, the irritation or itching of the hands of workers is considered due to the protease action of the hepatopancreas of prawns. According to some research reports, this protease of prawns contains collagenase which decomposes collagen. Thus the earlier assumption appears to be acceptable.

Euphausia rapidly lose freshness upon being caught. This is because of self-digestion (autolysis) due to their high protease activity. However, detailed investigations have revealed that the protease activity of *Euphausia* per se is not at all high and that the mechanical breakdown of the systems due to weakness of the *Euphausia* body results in rapid autolysis.

Euphausia caught with prawns and crabs blacken severely upon atmospheric exposure. As in the case of cut fruits, the blackening or browning is thought to be due to the action of the enzyme tyrosinase. This is controlled to some extent when treated with sulfite. In the case of prawns and crabs, the capacity to ozidize phenols, which like tyrosine have one —OH, is negligible or absent. Hence it is not the same as in the case of fruits. Instead, as mentioned earlier, they exhibit strong enzymatic activity to phenols with two —OH. It is not known, however, what exactly these phenol compounds are on which enzymatic activity occurs when the prawns turn black.

REFERENCES

Association of Chemistry, Japan (Nihon Kagakkai) (ed.). 1979. Elements of Chemistry, no. 25. Chemistry of Natural Marine Products. Gakkai Publ. Center.

Association of Fisheries, Japan (Nihon Suisan Gakkai) (ed.). 1977. Marine Biology Series no. 20. Fishmeat Protein. Koseishakoleikaku.

Association of Fisheries, Japan (Nihon Suisan Gakkai) (ed.). 1978. Marine Biology Series no. 22. Fish Culture and Fat Content in the Feed. Koseishakoleikaku.

Association of Fisheries, Japan (Nihon Suisan Gakkai) (ed.). 1979. Marine Biology Series no. 25. Carotenoids of Marine Organisms. Koseishakoseikaku.

Association of Fisheries, Japan (Nihon Suisan Gakkai) (ed.). 1979. Marine Biology Series no. 27. Biochemical Resources of Seas and Oceans. Koseishakoseikaku.

Association of Fisheries, Japan (Nihon Suisan Gakkai) (ed.). 1979. Marine Biology Series no. 29. Marine Food Products. Koseishakoseikaku.

Association of Fisheries, Japan (Nihon Suisan Gakkai) (ed.). 1982. Marine Biology Series no. 40. Glyceride-free Fat of Marine Organisms. Koseishakoseikaku.

Association of Resources Survey, Dept. Science and Technology. 1982. Standard Components of Japanese Food. Ministry of Finance.

Ikeda S (ed.). 1981. Minute Contents of Fishes and Mollusks. Koseishakoseikaku.

Shiino K. 1969. Biology of Marine Invertebrates. Baifukan.

Waterman TH (ed.). 1960, 1961. The Physiology of Crustacea. Academic Press, London—NY, vols. 1 and 2.

Other references in the text were taken from the Journal of Fisheries Association of Japan, Seas and Organisms, Nutrition and Food, and Journal of Nutrition Science.

7

Utilization and Processing of Prawns and Shrimps

The characteristic flavor of prawns and shrimps is liked by people through-out the world. Japanese people in particular are very fond of crustaceans and, in fact, Japan consumes 10% of the shrimp catch made in various countries. The nature of shrimp flesh differs considerably from that of fish meat. Hence it is necessary to have a perfect knowledge of the spe-cial features of shrimps before utilizing or processing them.

In this Chapter opinons pertaining to the chemistry and processing techniques of shrimps are considered and various aspects of shrimps such as freshness, color change, food product hygiene, freezing, canning, cooking, and dressing as special products, and utilization and processing of euphausiids are studied.

7.1 SHRIMP FRESHNESS

7.1.1 Deteriorating Phenomenon Following the Catch

Following the catch, the condition of shrimps deteriorates due to en-zymes in the body and adhering microorganisms. Initially the fresh red color of the caudal section turns bluish-white. The shell turns greenish, brownish, then finally black. The shrimp loses its original hardness and elasticity and the joint between the exoskeleton and the caudal section loosens. The head becomes readily detachable. The odor of fresh shrimp gradually changes to that of ammonia or a foul odor. Unlike other inver-tebrates and vertebrates, shrimps have soft and weak connective tissue which is highly fragile. In *Pandalus borealis* the flesh contains 75-80% water and the body fluid has salts, sugars, nucleotides, trimethylamine oxide (TMAO), and various amino acids. Therefore it is ideal for the propagation of bacteria and readily decomposes.

7.1.2 Deterioration of Freshness

Several methods have been suggested to ascertain the degree of freshness of fishery products including prawns. These can be broadly classified into judgment by the five senses, physical judgment, microbiological judgment, and chemical judgment. In the chemical judgment method, widely applied, besides the K value there are other indicators for freshness such as volatile base-nitrogen or VB-N and trimethylamine or TMA.

Figure 7.1 shows the changes in the volatile base-nitrogen, trimethylamine-N, and hydrogen ion concentration (pH) when shrimp are preserved at 5° C. The volatile base-nitrogen contains ammonia formed by bacterial action and volatile amines. Trimethlyamine is formed from trimethylamine oxide present in the flesh by the reducing enzymes of bacteria. The quantity of volatile base-nitrogen gradually increases and attains 30 mg/100 g after 4 days and exceeds 80 mg/100 g after 7 days. At this stage the shrimp decompose. Trimethylamine-N is 3.9 mg/100 g after 4 days but thereafter rapidly increases. Most fishery products are judged as follows: for 100 g of flesh, if the level of volatile base-nitrogen is 5-10 mg, it is adjudged fresh, 15-25 mg normal, 30-40 mg in the stage of early decomposition, and above 50 mg as distinctly decomposed. According to this standard *kuruma* shrimp preserved at 5° C start to putrify on the fourth day. Initially the common bacterial count is less than 300 and the colon bacteria are anaerobic, but after 4 days of preservation they increase to 460×10^3 and 20×10^2.

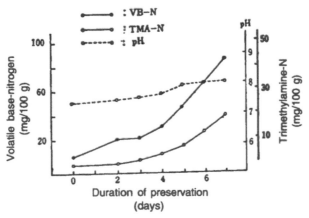

Fig. 7.1. Deterioration of freshness in shrimp preserved at 5° C (Uchiyama *et al.*, 1974).

When *kuruma* shrimp are preserved by freezing soon after killing by beheading, there is almost no change in the volatile base-nitrogen, trimethylamine-N, and pH. Even after long duration of preservation for 30 months, the volatile base-nitrogen is 15 mg/100 g and trimethylamine-N 0.8 mg/100 g (Fig. 7.2).

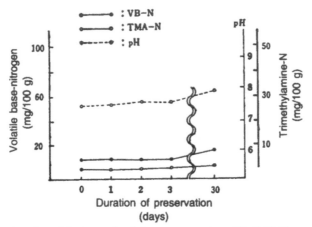

Fig. 7.2. Deterioration of freshness in shrimp preserved at –23° C (Uchiyama *et al.*, 1974).

7.1.3   Freshness Judgment Based on K Value

Adenosine triphosphate (ATP) undergoes enzymatic degradation in the muscles of fishes and shellfishes after their death, as shown in Fig. 7.3, and results in the final products called inosine (HxR) and hypoxanthin (Hx).

Changes in ATP traced through ion exchange chromatography are detailed in Fig. 7.4. In the case of instantly killed shrimp (Fig. 7.4, A), the ATP content is high while inosine, hypoxanthin, andinosinic acid (IMP) contents are low. When these shrimp are preserved at 10° C for 6 days, ATP, ADP, and AMP contents decrease as shown in (B) of the Figure. However HxR, Hx, and IMP contents increase and concomitantly the shrimp meat decomposes.

Judgment by the K value is the method used to assess the degree of freshness based on the degree of ATP decomposition or, in other words, the degree of accumulation of HxR and Hx.

The K value is determined with the following equation:

$$\text{K value (\%)} = \frac{\text{HxR} + \text{Hx}}{\text{ATP} + \text{ADP} + \text{AMP} + \text{IMP} + \text{HxR} + \text{Hx}} \times 100$$

In other words, the K value is the percentage of inosine and hypoxanthin against the total amount of ATP and the products of decomposition. The smaller the K value, the better the degree of freshness.

Uchiyama *et al.* studied the relationship between the K value and the degree of freshness for many fishes. The results indicated that instantly killed fishes have less than 10% and the high-grade *sushi* varieties (fishes best suited for the Japanese cuisine called *sushi*) about 20%. An example

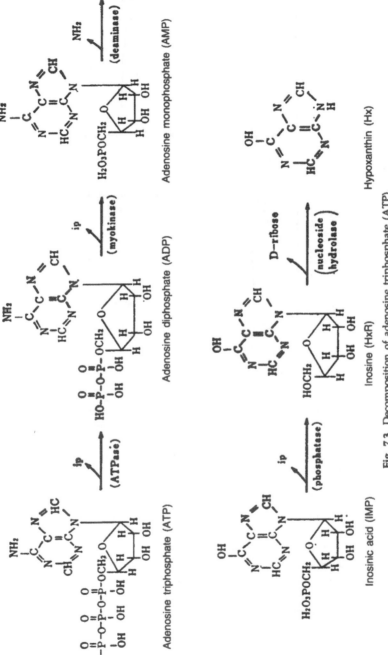

Fig. 7.3. Decomposition of adenosine triphosphate (ATP).

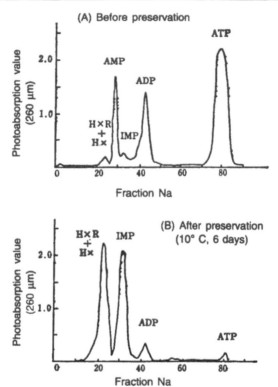

(A) Before preservation

(B) After preservation
(10° C, 6 days)

(A)   Ion exchange chromatography (resin: Dowex 1 × 4 Cl⁻) of ATP and its decomposed products in instantly killed shrimp
(B)   Ion exchange chromatography of ATP and its decomposed products of shrimp preserved at 10° C for 6 days

Fig. 7.4. Changes in adenosine triphosphate of shrimp meat preserved at about 10° C (Uchiyama *et al.*, 1974).

is shown in Fig. 7.5. While the volatile base-nitrogen content is applied as a parameter to judge the degree of freshness of fishes and shellfishes after the onset of the action of decomposing bacteria, the K value is applied as an effective parameter to judge the live condition. When the author examined the K value of commercial frozen shrimps, he found that most of them had 20-30%, the lowest was 8%, while the highest was 44%. The K value is measured by ion exchange chromatography, thin layer chromatography, or high-speed fluid chromatography.

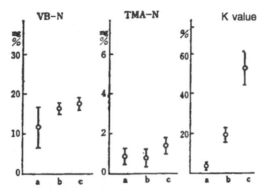

(a) Instantly killed fish
(b) High-grade horse mackerel and high-grade *sushi* types
(c) Common *sushi* types
⊢—O—⊣ indicates the mean values and their 95% reliable sections

Fig. 7.5. Comparison of freshness judgment based on VB-N, TMA-N, and K value (Uchiyama *et al.*, 1970).

7.1.4 Heat Stability of Protein

The heat stability of muscle protein, in particular the myofibril protein of marine fishes and shellfishes, depends on the type of fish and shellfish. Figure 7.6 shows the heat stability of fleshy prawns, American lobsters, horsehair crabs, Northern shrimps (*Pandalus borealis*) *Euphausia*, and mackerels with the changing speed of myofibril C-ATPase as the index. The vertical axis shows the changing speed constant (K_D) and the horizontal axis the temperature. The relationship between the log value of K_D and the heating temperature ($1/T$) is expressed by a regression straight line. Those which occupy the straight line more and more on the right side are unstable in heat. Between fleshy prawns and Northern shrimps, the changing temperature zone of the muscle protein of Northern shrimps is lower, thus indicating a lower heat stability similar to that of *Euphausia*.

The changes in protein solubility during preservation are shown in Fig. 7.7. In this study whiskered velvet shrimps (*Metapenaeopsis barbata*) were put in a blister sterol box and preserved at 5° C. Solubility of protein was measured periodically. The degree of freshness was assessed on the basis of pH, volatile base-nitrogen, and degree of blackening. The third day of preservation was concluded as the initial period of decomposition. As decomposition progresses, the meat becomes soft and emits a foul odor. In this case, when the deterioration of freshness became severe after 5-7 days, the solubility of salt-soluble protein was almost the same as that of shrimps in fresh condition. Since the fluid double refractibility changed, it was not possible to actually observe the

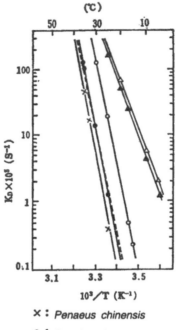

(℃)

X : *Penaeus chinensis*

● : *Panulirus hommarus*

— : Mackerel

○ : Horsehair crab

▲ : *Euphausia*

△ : *Pandalus borealis*

Fig. 7.6. Heat stability of crustacean myofibril protein (Hashimoto *et al.*, 1982).

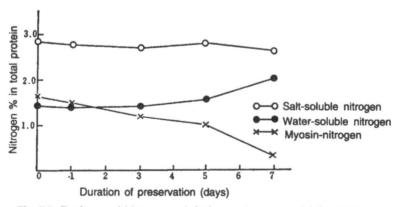

○—○ Salt-soluble nitrogen

●—● Water-soluble nitrogen

×—× Myosin-nitrogen

Fig. 7.7. Freshness of *Metapenaeopsis barbata* and protein solubility (Seki *et al.*, 1980).

phenomenon. Hence it was concluded that the protein undergoes qualitative changes along with decline in freshness. The water-soluble protein showed an increasing trend while the myosin protein gradually decreased. After 3 days of cold storage the protein completely loses its capacity to form gel.

7.2 COLOR CHANGES IN SHRIMP

7.2.1 Whitening

When small shrimps such as *Metapenaeopsis barbata* are dried they turn white. Whitening occurs when frozen shrimps are dried after defrosting or when fresh shrimps are dried without freezing.

It is not because of the fading red pigment astaxanthin that whitening occurs, but rather the white opaque substance which is soluble in acid and insoluble in water gets refracted between the cells containing astaxanthin and the exoskeleton. When the whitened part is observed under the microscope, innumerable spherical bodies are found. When HCl is added to these bodies, they become frothy and gradually disappear. Hence this material is assumed to be calcium carbonate. Details regarding the mechanism of ejection of calcium carbonate are not yet known. Perhaps it is due to the decrease in water content.

As a countermeasure against whitening it is effective to treat shrimps with sodium erythropinic acid or ascorbic acid.

7.2.2 Blackening and Its Prevention

(1) Blackening due to enzymes

When shrimp are kept without treatment after catching, they start to blacken within a few hours. Blackening occurs rapidly in the head, followed by the swimming legs, thoracic appendages, and exoskeleton. Muscular tissues do not turn black. Since blackening appears much earlier than decline in freshness of the flesh, it is not considered related to deterioration in freshness. When the body fluid and the liver of spiny lobster are left mixed together, blackening occurs within a short period. Therefore blackening is attributed to the presence of polyphenoloxidase. But as mentioned in the last chapter (6.6.2), details are not yet clear.

(2) Prevention by sulfite treatment and associated problems

Prevention of blackening of shrimps can be effected either by rendering the oxidizing enzyme of phenols with two —OH inactive or by avoiding phenol compounds which serve as the base. Inactivation of the enzyme by heat treatment prevents blackening but the product loses value as fresh shrimps. Hence low temperature treatment is applied to control enzymatic action as much as possible. All these treatments have certain limitations in prevention of blackening. Considering the significance of inactivating enzymatic action, it seems that proper adjustment of pH with organic acid could also prevent blackening effectively. In reality, since the neutrality of shrimp meat is strong, pH adjustment is difficult.

These days sodium sulfite salts, in particular sodium metabisulfite is said to be very effective in preventing blackening. Therefore, as a pre-treatment to freezing shrimps, chemical treatment is carried out with sulfites as the main agent. However, Japan has strict regulations relating to food and the standards for using such agents are specified. The residual amount in the case of peeled shrimps should be below 100 ppm for sulfur dioxide (SO_2). Blackening can also be prevented by immersion treatment in 0.5% solution of sodium erythropinic acid.

Table 7.1 shows the results of studying the effect of sodium metabisulfite to prevent blackening. *Kuruma* shrimps immersed in 0.3, 0.5, and 0.7% sodium metabisulfite solution were compared with shrimps immersed in sea water for 20-30 seconds. Observations revealed that blackening can be prevented to a great extent by the immersion treatment in 0.7% sodium metabisulfite for 10 minutes but the residual SO_2 is more than 100 ppm, i.e., it exceeds the standard specifications. Therefore it has become common practice to carry out immersion treatment in 0.3% solution for 10 minutes. If the duration of immersion is prolonged, the flavor is affected.

Table 7.1: Effect of sodium metabisulfite in preventing blackening (Tawara, 1972)

Treatment method	Duration				Remarks
	5	24	48	72	
Control	+~+	+~#	#	#	Blackening indicated by codes
0.3% immersion, then washed	—+	±~+	+~#	+~#	— No change
0.3% immersion	—	—~+	+~±	+~#	± Slight appearance of black spots
0.5% immersion, then washed	—	—~+	—~+	±~+	+ Tiny black spots appear on tail and abdomen
0.5% immersion	—	—~+	—~+	—~+	# Spreading of black spots
0.7% immersion, then washed	—	—~+	—~+	—~+	# Wide spreading of black spots
0.7% immersion	—	—	—	—~±	# Entire body appears black

Codes with thick letter indicate black-ening in most parts

Note: Duration of immersion in $NaHSO_3$ 10 min and duration of washing 20-30 s. Temperature during preservation, 2-5°C

A comparison of heat treatment and residual SO_2 is presented in Table 7.2. Sodium metabisulfite added to the shrimps decreases when heat treatment has been done. The decreasing rate depends on heat duration and salt concentration. In most cases, about 50% is left after heat treatment.

Table 7.2: Relationship between heat treatment and residual SO_2 in shrimps (Tsukuda *et al.*, 1972)

Types		Headless but with exoskeleton			Peeled
Body parts		Exoskeleton	Muscles	Entire body	Entire body
75° C	0	148 ppm	34.6	54.0	63.4 ppm
	2.5	109	23.7	39.5	43.1
		(73.6)*	(60.5)	(73.1)	(68.0)*
Duration	5	105	19.4	35.1	24.8
of heating		(70.9)	(56.1)	(65.0)	(39.1)
(min)	10	87.0	16.5	29.4	24.0
		(58.8)	(47.0)	(54.4)	(37.9)
100° C	0	315	84.0	123	147
Duration	1.5	246	68.1	98.3	117
of heating		(78.1)	(61.0)	(79.9)	(79.6)
(min)	3	123	43.5	57.0	75.0
		(39.0)	(51.8)	(46.3)	(51.0)

* ( ) indicate % before heating.

Makino *et al.*, (1977) measured SO_2 content and formaldehyde (HCHO) content of commercial shrimps. The results are given in Table 7.3. On the other hand, Yamanaka *et al.*, (1977) measured the formation of formaldehyde for a certain period when *kuruma* shrimps were separated into flesh and exoskeleton and immersed in 0.5% and 5% sodium metabisulfite. The results are given in Fig. 7.8. It is evident from the Figure that the

Table 7.3: SO_2 content and formaldehyde content in commercial shrimps (Makino *et al.*, 1977)

Type of specimen examined (native place)	SO_2 (g/kg)	HCHO (ppm)
Shrimp with shell (Madagascar)	0.123	28.8
Shrimp with shell (Indonesia)	0.054	27.4
Shrimp with shell (Nigeria)	0.133	13.9
Shrimp with shell (Madagascar)	0.054	15.1
Shrimp with shell (Nigeria)	ND*	ND
Shrimp with shell (Australia)	0.147	47.2
Shrimp with shell (Madagascar)	ND	5.0
Peeled shrimp	ND	3.0
Peeled shrimp	ND	ND

*ND: Shrimps in which SO_2 and HCHO were not detected.

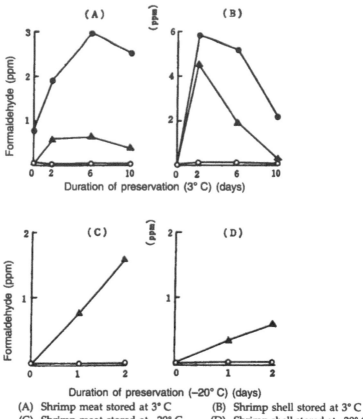

(A) Shrimp meat stored at 3° C   (B) Shrimp shell stored at 3° C
(C) Shrimp meat stored at –20° C   (D) Shrimp shell stored at –20° C
O: Control (untreated);   ▲: 0.5% treated;   ●: 5% treated

Fig. 7.8. Formation of formaldehyde due to sodium metabisulfite treatment (Yamanaka *et al.*, 1977).

shrimps not immersed in sodium metabisulfite solution did not form formaldehyde when preserved in 3° C and frozen at –20° C. However, when treated with 0.5% sodium metabisulfite for 5 minutes, formaldehyde was detected in the muscles at 0.6 ppm 2 days after cold storage and at 1.5 ppm 2 minutes after freezing. Formation of formaldehyde in the muscles of shrimps appears to be due to enzyme-free action from trimethylamine oxide as a result of sodium sulfite treatment. This formaldehyde decreases considerably on washing with water following treatment with sodium sulfite solution.

(3) Blackening during canning

Blackening which occurs in canned shrimp is not due to enzymatic action. Either during the heating to kill the bacteria or when the starting temperature is high after antiseptic treatment, the hydrogen sulfide formed from the sulfur compounds contained in the muscles, or when the water-soluble sulfides react with tin or iron of the canning material, black sediments form on the can surface (the main components are SnS and FeS). The mechanism of formation inside the can is reportedly that shown in Fig. 7.9. The reaction is influenced by pH, temperature, and pressure.

Cathode: $Fe \rightleftarrows Fe^{2+} + 2e$, $\quad Sn \rightleftarrows Sn^{2+} + 2e$

Anode: $\quad 1/2\ O_2 + 2e + H_2O \rightleftarrows 2OH^-$

Food product: $R - SH + SH - R \xrightleftharpoons[\text{Reduction}]{\text{Oxidation}} R - S - S - R$

$$\updownarrow$$

$$S^{2-} + R - H$$

$$\updownarrow$$

$$HS^- \rightleftarrows H_2S \uparrow$$

Solution: $Fe^{2+} + S^{2-} \longrightarrow FeS$ (black)

$\qquad Sn^{2+} + S^{2-} \longrightarrow SnS$ (brown)

$\qquad Fe^{2+} + 2OH^- \longrightarrow Fe(OH)_2 \xrightarrow{\text{Oxidation}}$

$\qquad Fe(OH)_2 \cdot Fe(OH)_3 \longrightarrow Fe(OH)_3$
$\qquad$ (black) $\qquad\qquad$ (red)

Fig. 7.9. Mechanism of formation of blackening substance in canning (Ota, 1980).

7.3 FOOD HYGIENE

7.3.1 Detection of Hygienic Condition

(1) Normal bacterial count

The normal bacterial count refers to the number of bacteria living in the food item. As the index of bacteriological quality of the food item, it is expressed in terms of the number of bacteria per gram of food.

The measured value of bacterial count differs according to the temperature of the culture. According to the food hygiene regulations, the temperature is set at 35° C. This indicates the presence of moderate temperature bacteria. When the number of moderate temperature bacteria is extremely large, this means that the food has spoiled due to improper handling or that it has been left out for a long period at room temperature whereby the propagation of moderate temperature bacteria is encouraged. Therefore, a large normal bacterial count in food indicates contamination by mixing and propagation of bacteria which can cause orally contagious diseases or food poisoning (most of the bacteria in this case are moderate temperature bacteria). At the same time, a low bacterial count cannot always be considered a sign of excellent quality of food in terms of bacteriology. For example, when shrimps infected with a large concentration of yellow bacterial clusters are heat treated, the bacteria themselves perish but the enterotoxin produced by the bacteria is heat resistant and therefore remains in the product and causes food poisoning.

(2) Intestinal bacteria

The bacteria causing infectious diseases of the digestive system, for example bacteria causing cholera, typhoid, dysentery, etc. are present in polluted food items. It is difficult to detect these pathogenic bacteria beforehand. Therefore assessment is based on checking the large bacterial groups already present in the intestine of human beings and animals. In other words, the presence of bacterial groups in the large intestine suggests the possibility of digestive organs being polluted by bacteria causing infectious diseases. This is a good index for judging the hygienic condition of food items. In the bacterial groups of the large intestine, there are gram-positive, aerobic, and anaerobic bacteria which decompose milk and sugar, and produce acid and gas.

7.3.2 Cooked and Frozen Food and Their Contamination

The cooked and frozen food prescribed by the Agriculture and Forestry Standards of Japan (No. 155) include shrimp fry, shrimp wafers, springroll, hamburger steak, meatballs, fish hamburger and fishballs. Sometimes a considerable quantity of normal bacteria and large intestinal bacteria is detected. Figure 7.10 gives the results of surveying the actual conditions of bacterial contamination during the production of shrimp fry.[1] The results indicate that they accord with the prescriptions for frozen food promulgated by the Ministry of Health and that there is no problem relating to bacterial infections. Bacterial infections occur due to secondary pollution during processing and also contamination of imported frozen shrimps at the point of origin.

7.3.3 Disinfection by Gamma Irradiation

In shrimps, irradiation is done with gamma rays of Cobalt 60 or electron rays (4 MeV) to kill the germs. This ensures a long period of preservation. In 1970, radurization[2] was developed and practiced in The Netherlands. In 1976, the products were put on test sale. In Japan, irradiation of shrimp is not allowed.

---

[1]It should be less than 50%. However, for shrimps measuring less than 6 g in body weight (shell removed), it may be less than 60%.
[2]Radurization refers to irradiation of 0.5-10 kGy (0.05-1 Mrad) for the purpose of reducing decomposing bacteria. In The Netherlands, 1 kGy is used. One Gy refers to the unit of absorption rays equivalent to 10^4 erg of energy for 1 g of material.

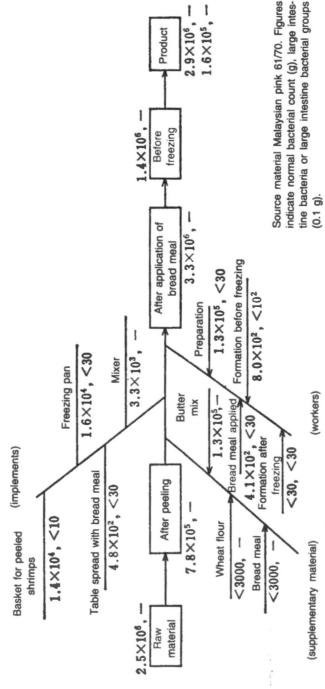

Fig. 7.10. Bacterial count and production steps in shrimp fry (Oyama et al., 1975).

The terms used in connection with food irradiation are radappertization (complete disinfection, 25-50 kGy), radicidation (irradiation of pathogenic germs, 310 kGy), and radurization (reduction of decomposing bacteria, 0.5-10 kGy). These terms became popular after the international symposium held in 1967 in West Germany.

Shrimps irradiated at 45-56 kGy and preserved for 7 months (0.5-1.5° C) were examined for their organic functions. In the evaluation of 9 points, irradiated shrimps scored 5.79 for odor, abnormal flavor, and taste as against the control group which scored 6.25. Regarding the preservability (Table 7.4), duration of preservation for nonirradiated shrimps was 14-16 days versus 21-28 days for the irradiated lot. When the shrimps were irradiated with gamma rays after heat treatment, they could be preserved at 28-30° C for 6-8 weeks while the nonirradiated ones could be kept for only a few days (Table 7.5).

Table 7.4: Duration of preservation of irradiated food (Sato, 1969)

Name of product	Dose (kGy)	Preservation temp. after irradiation (°C)	Duration of preservation of irradiated products (days)	Duration of preservation of nonirradiated products (days)
Shrimps	0.15-0.25	0.5-1.5	21-28	14-16
Fish (haddock)	0.1-0.25	0-1.5	30-37	10-14
(cod)	0.1-0.25	0.5-1.5	30	12-14
Clam	0.15-0.25	0.5-5.5	20-30	2-8
Fish with red flesh	0.05-0.15	0.5-3.5	10-21	5-7

Table 7.5: Prolongation of preservation period of heat-treated shrimps by irradiation (Savagon *et al.*, 1972)

Heat treatment temp. (°C)	Duration of treatment (min)	Dose (kGy)	Duration of preservation of nonirradiated products (days)	Duration of preservation of irradiated products (days)
100	15	0.35	1-2	42-56
109 (5 atm)	20	0.20	2-5	42-56
121 (15 atm)	8	0.10	3-15	42-56

7.4 FREEZING OF SHRIMPS

7.4.1 Classification of Frozen Products

The frozen products of shrimps include fresh frozen in which shrimps are frozen fresh and cooked and frozen products in which shrimps are exposed to steam or hot water to coagulate the protein and then frozen. The products of export and import are fresh frozen. In the case of spiny lobsters first heat treated at 90° C for 70 seconds and then frozen, shell removal from the flesh is easier when they are defrosted and cooked. This treatment is widely applied.

Again, based on the place where the freezing process is carried out, frozen products are classified as follows. When the catch is brought ashore and frozen in cold-storage plants, the products are termed land-frozen. On the other hand, when the trawlers have the requisite facilities for freezing and the catch is frozen on board, the products are labeled ship-frozen. Ship-frozen products are excellent in freshness but their production constitutes only 10% of the total frozen products.

7.4.2 Processing Forms

Shrimps are processed in four forms and then frozen.
1. *With head (round, whole, head attached)*: This refers to the whole shrimp. In this case, large shrimps with a high degree of freshness (tiger shrimp, white shrimp, etc.) are used.
2. *Headless but with legs or headless with shell or with tail flesh (also known as headless, shell on, head off, tail)*: This refers to the abdominal part with legs, telson, tail fan, and shell but without the cephalothorax. In this case large species such as *Penaeus monodon* are used.
3. *Headless and legless (also known as dressed)*: This refers to the abdominal part with tail fan and shell but without the cephalothorax and abdominal legs.
4. *Peeled, shelled shrimp*: This is further classified as follows:
    a. *Peeled:* The entire shell including that of the cephalothorax and tail fan is removed. This is further classified into three types.
        (i) PUD *(peeled but not deveined)*: This is the Japanese style wherein the dorsal part is left intact.
        (ii) PD *(peeled and deveined)*: This is the American style in which the dorsal part is removed.
        (iii) CP *(cooked and peeled)*: This is the Australian style. It is basically the PUD type but the shrimps are cooked.
    b. *Tail fan, peeled*: This is classified into two types: in one the shell of abdominal segment VI is retained, while in the other the shell is removed.

7.4.3 Freezing Procedure

The steps normally adopted in freezing shrimp are plotted in Fig. 7.11. First the head is removed and selection done. Shrimps with black spots, white spots, soft shell, and broken parts are excluded. The muscles between the head and the abdomen, popularly known as the triangular muscles, are carefully excised so as to improve the meat yield content. The efficiency of head removal per worker is 24-34 shrimps/minute provided the person is industrious.

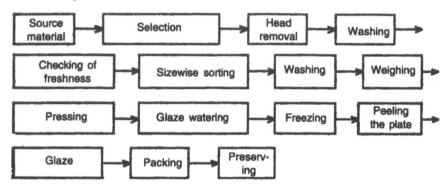

Fig. 7.11. Steps in freezing shrimps.

Next, the dressed shrimps are washed to remove any dirt. It is very essential to wash off hemocyanin, which is equivalent to the blood pigment of high animals. Hemocyanin is present in the body fluid of mollusks and arthropods. It is a carrier of oxygen. It is a copper protein and in due course of time turns bluish and loses the ability to flow. Eventually it solidifies and in this state cannot be washed off. Hemocyanin of arthropods is a polypeptide chain (molecular weight, about 75,000) containing two copper ions. Six chains make one unit of hemocyanin. The size of the molecules varies in arthropods from species to species.

The next step is to sort the shrimps according to size either manually or using a separator. After sorting, the shrimps are washed in water and weighed. At the time of weighing, the likely reduction in weight due to drying while freezing and preserving is taken into consideration and usually 3-5% is added to the weight. The cake arrangement in which the shrimps are pressed in a freezing pan, is of two types. In the first type, layer packing, the shrimps are systematically arranged on the freezing block. In the other type, they are packed at random, and this is known as random packing or jungle packing. In layer packing, either a double-sided or single-sided arrangement is done. When packing includes freezing the shrimps directly inside the carton, this is termed raw packing. Because the cake arrangement looks attractive, it is very popular in the packaging of frozen products.

In the next step, in the case of a frozen pan, sufficient water is injected to ensure total submersion of the shrimp. In row packing, water is so injected that the polyethylene sheet covers the surface of the shrimp and the lid of the inner carton. Thereafter the steps include freezing, pan removal, and packing. Pan removal after freezing is done either by immersing the frozen pan in a deep freezing tank or by continuous exposure to a shower. The pan-removed product is called the cake or block. Now the glaze is applied and the product placed in a polyethylene bag. Then it is packed in an inner carton of waterproof cardboard (priorly coated with wax). Glaze treatment is carried out after freezing to prevent desiccation inside the package.

While packaging, care is taken to minimize weight loss and to prevent changes in color, fat content, and flavor. Vacuum packing is recommended as it minimizes deterioration of product quality (Fig. 7.12). In this case a plastic film with low oxygen permeability is used. The main types of packaging used for commercial frozen food products are listed in Table 7.6.

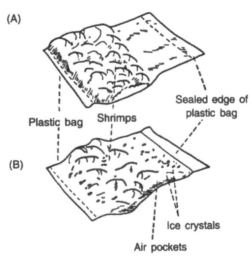

(A)   Vacuum packing. Oxidation and color change of shrimp effectively prevented.
(B)   Heat sealed (packing under normal pressure). The shrimps are exposed to oxygen inside the package. Thus they often become dry and ice crystals form inside the packet. In this type of packaging, the shrimp flesh loses its quality.

Fig. 7.12. Two types of packaging of frozen peeled shrimps (J. Aagaad, 1973).

Table 7.6: Packaging of fishery products and cooked food items

	Items	Packing components
Fisheries	Peeled shrimps	PE
	Tempura shrimps	ON/PE (Ps flat sheet)
	Sashimi squid	ON/PE (Ps flat sheet)
	Sardine	PE
	Crab	Contracted PVC
	Abalone	ON/PE
Cooked food	Croquette	OPP/PE, PET/PE
		MST/PE (PSP tray)
	Springroll	OPP/PE
		MST/PE (PSP tray)
	Baked	OPP/PE
		MST/PE (PSP tray)
	Pizza	Paper box
	Hamburger	ON/PE, ON/PE (PSP tray), paper box

Note: PE: polyethylene; ON: stretch nylon; PVC: polyvinyl chloride; PET: polyester; OPP: stretchable polypropylene; MST: moisture-prevention cellophane; PSP: polyesterol.

7.4.4 Freezing Equipment (Plant)

The equipment used for freezing is classified into the following types on the basis of heat enthalpy. Plate-type or airblast freezing equipment is generally used for shrimp (Table 7.7). In the case of plate-type freezing, the shrimp are placed in metal trays, pressed between metal plates (15-20), and frozen in brine. The duration of freezing depends on the water temperature used for glazing and the type of shrimp. If the temperature of freezing is –35 to –40° C, the duration is 4-6 hours. In the case of airblast freezing, the duration is 8-12 hours when the temperature is –15 to –18° C. Airblast freezing may be of the tunnel type, belt type, or fluid flow type. Other types of freezing are also practiced. For example, the packed shrimps may be frozen by immersion (brine), liquid nitrogen (LN), or liquid furon. Shrimps are suitable for IQF. products (individual quick frozen food).

Table 7.7: Medium of heat transfer and freezing equipment

Medium	Freezing equipment
Metal	Plate-type
Air	Airblast
Fluid	Immersion
Gaseous fluid	LN and furon

7.4.5   Relationship between Freezing Speed and Ice Crystal Formation

The freezing speed of shrimp differs according to the freezing medium and the type of freezing equipment. Accordingly the rapidity of formation of ice crystals in the shrimp meat differs in slow freezing and rapid freezing depending on the temperature of preservation.

Figure 7.13 shows the freezing curve when lobster is frozen at $-20°$ C and $-30°$ C. The slanting curve represents the zone of maximum ice crystal formation. Since the quantity of heat required (79.7 cal/g ice) is released from the shrimps when cooling is carried out under constant conditions, the cooling speed is slow. Due to this decrease in freezing speed, it takes time to pass the zone of maximum ice crystal formation and the ice crystals tend to become large in shrimp meat.

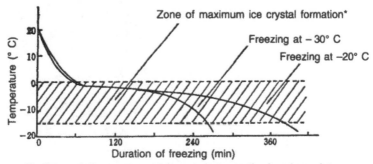

Fig. 7.13. Freezing curve of lobster (Suzuki *et al.*, 1968).

As for the temperature and the conditions of ice crystal formation, it is said that if the freezing temperature is low, many crystal sites will form in the meat and, furthermore, if the speed is slow, minute crystals will be abundant. On the other hand, if the freezing temperature is high, the sites of crystal formation in the meat are relatively fewer, their growth speed higher, and therefore relatively fewer but larger crystals will form. Figure 7.14 presents microphotographs of lobster meat frozen at $-30°$ C (B) and $-20°$ C (C) and preserved for six months. In the case of (B) where the freezing temperature was low, numerous small ice crystals are visible in the cells. In (C), where the freezing and preserving temperatures were higher than in (B), relatively large but fewer crystals can be seen inside and outside the cells. (A) in the Figure is a picture of raw lobster meat before freezing.

The degree of ice crystallization in shrimp meat, that is the freezing percentage, can be determined by the equation given below. In the case

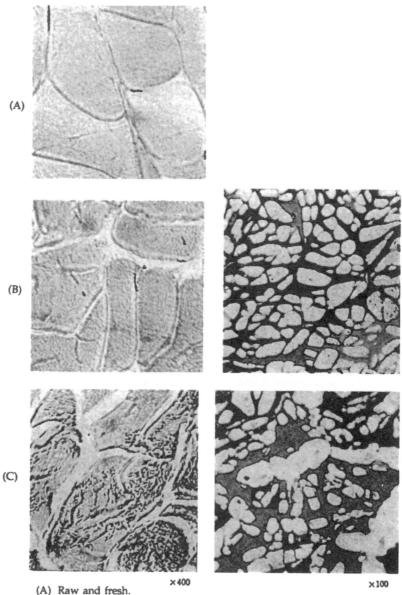

(A) Raw and fresh.
(B) Frozen at –30° C and preserved for six months. Some cracks and ice crystals appear inside the cells.
(C) Frozen at –20° C and preserved for six months. Many cracks and wrinkled cells appear. Ice crystals are large and present both inside and outside the cells.

Fig. 7.14. Microphotos of frozen lobster meat (Suzuki *et al.*, 1968).

of shrimps for which the freezing point is –2° C, if freezing is carried up to –20° C, then

$$\text{Freezing percent (\%)} = (1 - \theta_f/\theta_m) \times 100$$

where θ_f is the temperature of freezing point and θ_m the mean body temperature after freezing.

Thus 90% of the water becomes converted into ice crystals $(1 - (-2)/(-20) \times 100)$.

7.4.6 Size and Use

The sizing for shrimps is prescribed according to the international commodity specifications. Table 7.8 presents the specifications for *kuruma* shrimp. The code 11/15 refers to the number of shrimps per pound (1 pound = 453.69 g) and the number of shrimps in this case is 11-15. The same unit is applied for headless shrimps with legs, shrimps with head, and peeled shrimps also.

Table 7.8: Size specifications and use (Sakamuki, 1979)

Size specifications	Number/pound	Weight/specimen	Use* (Remarks)
Under 10	Less than 10	More than 43	Fry (mainly 4/6)
11/15	11/15	43-29	
16/20	16-20	29-22	Tempura (mainly 21/25)
21/15	21/25	22-18	
26/30	26-30	18-15	
31/35	31-35	15-13	*Sushi* (mainly 21/25, 31/35)
36/40	36-40	13-11	
41/50	41-50	11-9	Shrimp with bread meal
Over 51	More than 51	Less than 9	(mainly 51/60)

*Others, 60/80 fried tempura, 200/300 au gratin use.

The sizewise use of shrimps in Japan is mainly as follows. *Penaeus chinensis* and *P. monodon* with less than 15 count are used for fry. *Kuruma shrimp* of 21/30 count is used for tempura (fried in oil); *duruma* shrimp or Mexican brown shrimp of 21/50 count, which become red after cooking, are used in *sushi* (Japanese style raw meat preparation). Shrimps sold in supermarkets are mostly of the white variety at 31/60, average 41/50 count. Shrimps of India and Southeast Asia are 41/90 with most at 51/60 count. These are used for frozen shrimp mixed with bread meal. Peeled shrimps are mostly used in Chinese style cooking and fall in the category PUD.

7.5 CANNED SHRIMP

Production of canned shrimp in Japan has advanced during the past few years and the annual production in 1980 was 71 tons. *Pandalus borealis* or other imported frozen shrimps are used for this purpose. Production of the canned product includes treatment of the raw material, cooking, consolidation of the meat, removal of air, airtight packing (double packing) disinfection by heat, and freezing. Figure 7.15 outlines the procedure (wet packing with salt) involved in canning shrimp.

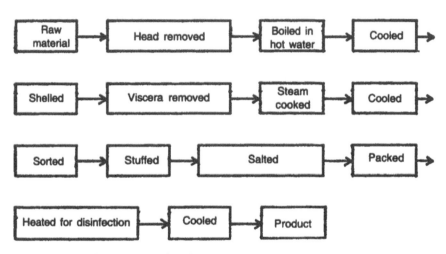

Fig. 7.15. Procedure for manufacturing canned shrimp.

If handling of the raw shrimp is inept, the digestive fluid is exuded and the meat decomposes. Therefore the shrimp should be thoroughly washed and compressed. Those which change color to yellowish-green as a result of disintegration of the liver or in which the cephalothorax blackens due to polyphenol oxidation are unsuitable for preparation of canned food. A peeling machine is used to remove the shell. However, the red pigment on the meat surface is lost and hence the color fades; consequently the yield declines. Therefore, peeled shrimps should first be dipped in acidic fluid (500 g acetic acid in 45 liters of water) for 15-30 minutes and then cooked. If the water is not completely removed, the contents often adhere to one another. For meat filling, the method given in Table 7.9 is presently used. In the case of dry packing, salt is not added after filling the meat. As shown in Table 7.10, up to 60-80% of the thiamine, niacin, pyridoxine, pantothenic acid, and folic acid in the shrimp meat are lost; only riboflavin is retained.

Table 7.9: Specifications for shrimp canning (Canning Material, 1976)

Item	Code mark	Can type	Solid quantity (g)	Conditions
Shrimps	PNN	Crab no. 1	250	Inner surface of can coated
		Crab no. 2	125	Contents wrapped in sulfuric acid paper
Small shrimps	SRN	Crab no. 2	150	Same as above
Small shrimps (broken ones discarded)	SRL	Crab no. 3	75	Inner surface of can coated Size should be uniform
In brine		Tuna no. 2	125	Fluid should be well above contents
Small shrimps (broken)	SRLH	Tuna no. 2	127	Inner surface of can coated
In brine				Fluid should be well above contents. Size should be more than half of original and can embossed with the code H.

Table 7.10: Loss of vitamins due to heat treatment (Thomas *et al.*, 1961)

Treatment	Thiamine	Riboflavin	Niacin	Pyridoxine	Pantothenic acid	Folic acid
Frozen peeled raw shrimp	18 µg	19 µg	0.9 mg	80 µg	200 µg	3 µg
Heated for 4 min in hot water, frozen and dried	9	19	0.4	64	40	1.5
Immersed in cold water for 15 min	9	—	0.3	—	—	—
Blanched in hot salt water for 10 min and canned; then heated at 116° C for 25 min	3.6 (80%)*	19 (0%)	0.2 (80%)	32 (60%)	40 (80%)	1.2 (60%)

*Loss percentage   (in 100 g)

7.6 COOKING AND SPECIAL PRODUCTS

7.6.1 *Tsukadani*

For the preparation of shrimp *tsukadani*, mainly boiled and dried small shrimps are used. These shrimps are pretreated by washing in hot water about 40° C for 3-4 min. The shrimps are then drained in a metal colan-

Fig. 7.16. Natural drying of boiled *sakura* shrimp.

der. In another method the pretreatment consists of washing with 1.5% acetic acid solution for 30 min and then draining thoroughly. The pretreated material is then added to the cooking medium (water 27%, sugar 9.3%, agar agar 0.1%, salt 4%, gluten 59.6%) already simmering in a vessel, allowed to cook for 30-40 min, then spread on a tray to cool. Cooking is hastened by means of fans.

7.6.2 *Fukashi* Shrimp (Dried Shrimp)

Relatively small shrimps are used for this culinary delight. It can be made as a raw dried product or cooked and dried product. Usually three types of shrimp are used: shrimp with shell, peeled shrimp, and minced shrimp. The rate yield is 36-43%, 32-39%, and 23-25% respectively. Shrimp with shell is used for cooking while the other two are used for roasting and Chinese-style recipes.

For *sakura* shrimp, so popular in Shizuoka prefecture, the shrimps are washed and spread on a black nylon net for drying in the sun (Fig. 7.16). This preparation is known as natural dried *sakura-ebi*. In another preparation the shrimp is cooked for 20-30 s in boiling salt water to which red coloring has been added, and then dried. This is called cooked and dried *sakura* shrimp (peeled). Other recipes are *kama-age sakura-ebi* in which the shrimp is cooked in a pot containing salt water with permissible coloring added, and immediately cooled; in *kimuki-ebi* the peeled shrimp is cooked in salt water, then dried.

7.6.3   *Ebisenbei* (Shrimp Wafers)

Popular in Kagawa prefecture. Fresh velvet shrimp (*Metapenaeopsis barbata*) is used for this preparation of unique flavor. After washing, the shrimps are deshelled and salt, glutamic acid, and red coloring added to achieve an attractive color and taste. The shrimp are then baked for 3 min in a wafer baking machine, after which they are dried for an hour at about 50° C.

7.6.4   *Kikuzuke* (Shrimp Pickle)

A special pickle product of Okayama prefecture. *Metapenaeopsis barbata* (velvet shrimp) is again the preferred shrimp species. Raw shrimp are peeled and kept in a 10-15% saline solution for about 1 week, after which a little sodium glutamate and soyabean paste one added and the shrimps subsequently transferred to rice malt. The ratio of malt to salted shrimps is almost equal. The material is then pressed with a heavy stone and stored for about 1 month before serving. However, the product has a much better taste after 2 or 3 months. Some people prefer a slightly sweet taste and hence sweet pickles are also made, but their shelf life is poor.

7.6.5   Shrimp Cake

Shrimps from the Tohoku region are used for this preparation. Raw shrimps cooked in salt water and a special rice used in the preparation of rice cake (*machi*) are mixed and pounded. The cakes are baked before eating. The combination of rice and shrimp flavors results in a delectable dish. In Nagasaki, the cakes are prepared from the flesh of spiny lobsters and fried in oil.

7.6.6   Other Special Products

There are several other special items prepared with shrimps in various parts of Japan. For example, *nabe-age* of Kanto Tsuchiura region, *aminoshitsuke* of Okayama, *fukashi* shrimp of Ube and shrimp-*kombu* of Shimono seki are very popular.

Outside Japan also, shrimps are used for various preparations, for example, *kapi* of Thailand and Malaysia, pickled small shrimp (Fig. 7.17 shows the method of preparations). In Indonesia, a dish called *kurupuku* is made with tapioca powder, a little salt, and minced rock shrimps. The mixture is then shaped into small rings on a skewer, dried in the sun, and fried in oil before serving.

7.6.7   Shrimp *Kamaboko* (Paste)

Efforts are underway to produce shrimp *kamaboko* using frozen minced shrimp. The steps for this culinary dish are shown in Fig. 7.18. First, natural coloring is added to the frozen minced meat. The main raw

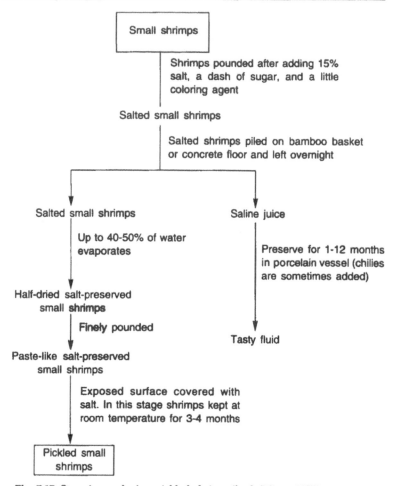

Fig. 7.17. Steps in producing pickled shrimp (*kapi*) (Miwa, 1978).

material, of typical shrimp flavor and form, is prepared. The exoskeleton material and the main raw material are suitably combined in a blender to form *kamaboko* shrimp. The *kamaboko* is heated for a short time in an ultraviolet or microwave oven and 65% water retained. It can be kept for a long time with taste continuously improving.

7.6.8 Shrimp Cooking

Among the raw shrimp preparations there are the peeled *hokkakuakaebi* (*Pandalus borealis*), also known as *amaebi* and *kurumaebi* (*Penaeus japonicus*) and the boat-shaped preparation of *isebi* (*Panulirus japonicus*) in which the hard shell is lifted and inserted into the head. In Japan, this is popularly known as *funenari*, meaning boat-shaped, or *odoriebi*, meaning dancing shrimp.

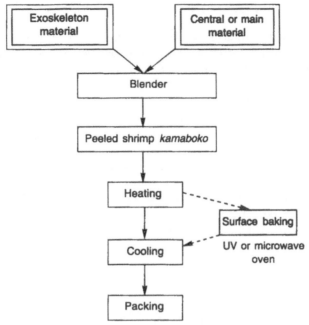

Fig. 7.18. Steps in preparation of shrimp *kamaboko* (Ishida, 1982).

Shrimps are cooked in various ways. They are boiled, baked, fried and even roasted. At wedding banquets, spiny lobsters soaked in the stock formed after boiling along with a little wine and salt are served. *Kuruma* shrimp are very popular as hors d'oeuvres, i.e., shrimp cooked in grape wine, rice wine, and a little salt and served with a dip. The preparation per se is known as cocktail-sauce dressed shrimp. If 10% sugarbeet gratings are added to this and the mixture made into small balls, the preparation after cooking is known as shrimp balls.

Among the boiled preparations there is the *gusoku* boiling or cooking in which the shrimp is cooked rapidly in the shell. The term *gusoku* means a soldier's armor.

There are also various ways of baking shrimp. Spiny lobsters, *kuruma* shrimp, *shiba* shrimp, *kawa* shrimp and others are popularly used and baked either in or without the shell.

Shrimp preparation also includes fried forms. In Japan, they are known as tempura and fried prawns. For tempura, the shrimp should be one- or two-year-old *kuruma* shrimp (18-35 g).

Among the dehydrated preparations, peeled shrimp as well as shrimp mixed with green peas are used. The dried shrimp is placed in water and left until the flavor is restored.

7.7 USES AND PROCESSING OF KRILLS (*Euphausia*)

Since 1972 much emphasis has been given to the effective use of krill resources. Various researches have been conducted. As of now, the demand for krills as a food product is low. Krills are mostly used as feed for other organisms.

As against shrimps in which 40-60% of the body is palatable, only 25-30% is edible in krills. A mixture of meat and shells reduces palatability. Moreover, the cephalothorax and the viscera contain proteases which are strongly active enzymes that degrade proteins (Fig. 7.19). Therefore, while treating, if by chance the cephalothorax is included, the enzymes are released, destroying the tissues, and causing subsequent blackening. The cephalothorax is easily crushed at a pressure of 40-140 kg cm^{-1}. Krill blackening can be controlled to a great extent by putting fresh krills in bags filled with nitrogen gas, which the enzymes cannot permeate (Table 7.11).

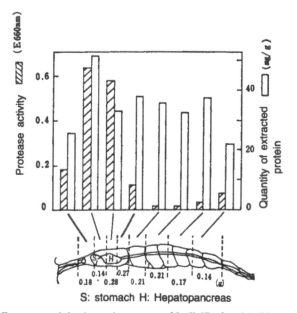

S: stomach H: Hepatopancreas

Fig. 7.19. Protease activity in various parts of krill (*Euphausia*) (Ohase, 1980).

The muscle protein of krill is not very stable against temperature and in this respect it resembles *Pandalus borealis* (see Fig. 7.6). To prevent degradation of krill meat protein, not only should preservation be done at low temperature, but an antiprotein-degradation agent such as sugars added. Krill meat has a strong tendency to absorb water. The water content of meat peeled manually is about 80% and in mechanized peeling about 90%.

Table 7.11: Blackening of krills when sealed in bags of various types of plastic film with differing permeability for enzymes (Gato, 1978)

Bag type		Experimental conditions			
Label[1]	Enzyme permeability (P_{O_2})[2] $(ml/m^2/24 \ h/atm)$	Kept undisturbed		Agitated	
		20° C 16 h	38° C 16 h	20° C 24 h	20° C 72 h
ON_{15} /$EVAL_{15}$ ®$_{15}$ /PE_{60}	0.9	—	—	—	—
ONP_{20}/$EVAL$®$_{15}$ /PE_{60}	1.0	—	—	—	—
PET_{12}/PE_{60}	157	—	—	—	+
OPP_{20}/PE_{60}	>2,000	+	+	+	++
PE_{100}	>2,000	+	+	+	++

[1] A product of Clare Company. The number indicates the thickness of the film (μm).

[2] 35° C, the value of relative temperature 0%. Frozen krills are placed in various types of bags (5 × 12 cm) which are then filled with nitrogen gas and sealed. The bags are kept at 20° C or 38° C. Those which show no blackening are marked —, those with slight blackening +, and those with significant blackening ++.

REFERENCES

Ota F. (ed.). 1980. Fisheries Processing Technique. Shinsuisangaku Zenshu. Koseisha-koseikaku, Tokyo, pp. 150-189.

Sakaguchi K. 1982. Science of Eating, no. 67. Special issue—Shrimp. Marunouchi Publ., Tokyo, pp. 52-78.

Sakamuki N. 1979. Shrimp. Suisansha, Tokyo, pp. 23-56.

Sato T. 1969. Radioisotope, 18 (10): 464.

Tsuda S., Tenno T. 1972. Research Report of Tokai Fisheries Research Center, 72: 9.

Uchiyama, Tenno, Kondo, Tanabe. 1974. J. Food Hygiene (Shokueishi), 15 (4): 301.

Ueno T. 1978. Japanese Cooking—Shrimp and Crab. Murata Publ., Tokyo, pp. 63-72.

8

Consumption and Circulation of Prawns and Shrimps

8.1 TRENDS IN SHRIMP CONSUMPTION

8.1.1 Trends in Quantity Consumed

The consumption of shrimps in Japan has risen steadily with greater westernization of food habits and the development of foreign food industries since World War II. High economic growth has also contributed to changes in food habits. Table 8.1 shows the quantity of shrimps consumed per household throughout all of Japan. In the first half of the decade beginning in 1965, it was 1.2-1.5 kg, which in the second half increased to 2.0-3.2 kg. These figures pertain to domestic consumption. If shrimp consumption outside the home were included, the actual increase in consumption would be 3-4 times greater.

Table 8.1: Changing trends in consumption of shrimps (per household throughout all of Japan)

Year	Consumption cost (yen)	Consumption quantitative (100 g)	Year	Consumption cost (yen)	Consumption quantitative (100 g)
1965	631	12.85	1974	4,347	31.98
1966	739	13.54	1975	4,910	31.14
1967	781	13.00	1976	5,224	29.06
1968	937	14.54	1977	5,360	26.55
1969	1,164	15.60	1978	6,104	29.49
1970	1,719	20.60	1979	6,283	27.68
1971	2,188	23.64	1980	6,593	27.48
1972	2,524	25.17	1981	6,860	27.50
1973	2,938	24.05			

Data: Statistical Division of Prime Minister's Secretariat, 1981. Data given for shrimps and crabs for 1965-1979 combined. Cost and quantitative figures for shrimps in 1980, estimated by the author from the totals given for shrimps and crabs.

In years past shrimp consumption was maximum in America. In 1965, Japan superseded America to occupy first place in shrimp consumption worldwide.[1]

During the period of economic recession the consumption rate neither increased nor declined. With 1974 as the peak period of consumption, domestic consumption thereafter held at 2.6-2.7 kg per household.

8.1.2   Trends in Consumption Cost

The cost of shrimps varies greatly according to the type. A comparison of the price per kilo of various major species sold in the central wholesale market of 10 major cities and towns of Japan in 1980, revealed that the cost of fresh spiny lobster was 6,320 yen, fresh *kuruma* shrimp 4,612 yen, frozen fleshy prawn (*Penaeus chinensis*) 2,113 yen, and fresh fleshy prawn 1,394 yen. All of these were of the best grade and sold in wholesale markets along with yellowtail, tuna, abalone, etc.

Both the consumption cost and quantity of shrimps consumed showed an increase that paralleled economic growth up to around 1975. From that year a sluggish trend set in. Table 8.2 reflects the depressed prices of frozen fleshy prawns, fresh fleshy prawns, and fresh *kuruma* shrimps handled in large quantitites in the central wholesale markets of 10 major cities and towns of Japan. Only fresh spiny lobsters, available in small quantities, continued to fetch a high price after 1975.

Table 8.2: Changing trend in price of shrimps and prawns (prices per kilo at wholesale markets of 10 major cities and towns)

	1976	1977	1978	1979	1980	(Quantity handled in 1980, tons)
Fresh spiny lobster	4,777 (100)	5,383 (113)	5,767 (121)	5,513 (115)	6,320 (132)	269
Fresh *kuruma* shrimp	4,794 (100)	4,659 (97)	4,080 (85)	4,036 (84)	4,162 (96)	1,009
Fresh fleshy prawns	1,873 (100)	1,820 (97)	1,575 (84)	1,665 (89)	1,394 (74)	1,751
Frozen fleshy prawns	2,238 (100)	2,279 (102)	1,987 (89)	2,508 (112)	2,113 (94)	6,872

Note: The ten locations include 6 major cities and the markets of Fushimi, Sendai, Hiroshima, and Fukuoka.

Data: Ministry of Agriculture, Forestry and Fisheries, 1980.

8.1.3   Characteristics of Consumption

(1) Business consumption

Shrimp are considered high-grade fishery products and there has been a striking increase in their business consumption. This is evident from day-

---

[1]In 1970, the consumption of shrimps per capita was 1.57 kg in Japan versus 1.10 kg in America (Sakamuki. Shrimp. Suisansha, November, 1979).

to-day life of the common people and it would be interesting to know the ratio of business consumption and household consumption in relation to total consumption. Unfortunately, no data which would give an accurate figure for the country are available. According to the data collected by a market survey carried out in Tokyo, business consumption covered 70-75% and domestic (household) consumption about 25-30%. Let us discuss this data.[2]

The survey carried out in 1971 collected data on total sales by 66 representative dealers selected from the imported shrimp wholesale dealers of the central wholesale market in Tokyo. According to this survey, of the sale of 536.3 tons, sales for business purposes constituted 402.1 and sales for domestic, 134.2 tons. This works out to a ratio of 75 : 25. Business mainly used these shrimp for the preparation of *sushi* Japanese style cuisine, Chinese cuisine, Western cuisine, and delivery services.

In 1978, a survey was carried out on the sale of important fishery products by 73 representative business establishments of Tokyo. According to the data of this survey, business use constituted 66%, business-cum-domestic 17%, and exclusively domestic 17%.

It is evident from the foregoing data that in major cities such as Tokyo and Osaka, there is an endless demand for shrimps for business use. In this context, it is interesting to study the condition of medium and small towns. The survey carried out in Toyohashi city with a population of 300,000 (1981) indicated that domestic consumption of shrimps constituted 51.2% and consumption outside the household 48.8% (Table 8.3). In

Table 8.3: Percentage of household and business consumption of fishery products (Toyohashi city)

Items	Household consumption			Business consumption				Grand total
	Nonagri-cultural household	Agri-cultural household	Total	Restau-rants	Social events	Vaca-tional resorts	Total	
Shrimp	46.8	4.4	51.2	29.3	8.4	11.1	48.8	100
Tuna	35.7	7.4	43.1	44.3	4.3	8.3	56.9	100
Mackerel	44.2	20.5	64.7	24.9	9.8	0.6	35.3	100
Horse mackerel	58.6	11.8	70.4	19.4	8.9	1.3	29.6	100
Pike	65.7	18.7	84.4	9.2	5.8	0.6	15.6	100
Yellowtail	50.7	6.5	57.2	36.3	2.9	3.6	42.8	100
Squid	56.4	7.5	63.9	25.7	5.9	4.5	36.1	100
Clam	68.4	—	68.4	23.2	3.2	5.2	31.6	100
Corbicula	83.3	—	83.3	8.2	7.6	0.9	16.7	100

Notes: 1. Comparison of consumption for one month.
       2. Items both fresh and frozen.
Source: Food Demand and Supply Research Center (1980).

---

[2]Also see "*sakamuke*" (shrimps).

other words, consumption by the two markets was roughly half and half. In large cities, on the other hand, business consumption was lower. However, together with tuna consumption of shrimps outside the home was higher. It differed significantly from in-house consumption of mackerel, horse mackerel, pike, squid, etc. which constituted 60-70%.

(2) Urban high-income household consumption

In domestic consumption, the difference in consumption cost is clearly evident between high-income and low-income households. As shown in Table 8.4, the difference in consumption between minimum income bracket (I) and maximum (V) for consumption of fishery products on the whole is about 1.7 versus 2.4 for shrimps.

Table 8.4: Shrimp consumption according to the five annual income brackets (per house hold, 1981)

Annual income brackets	Shrimp consumption	Bracket I taken as 100	
		Shrimp	All types of fishery products
I (265 × 10,000 yen)	4,248 yen	100	100
II (265-350 × 10,000)	5,515	129.8	114.7
III (350-444 × 10,000)	6,342	149.3	124.3
IV (444-588 × 10,000)	7,915	186.3	141.9
V (above 588 × 10,000)	10,281	242.0	171.6

Source: Statistical Division of Prime Minister's Secretariat, 1981.

When shrimp consumption was compared between the urban salaried household and the rural household, it was found to be 2.7 kg and 6,600 for the former versus 1.7 kg and 4,300 yen for the latter. The difference between the two households is significant.[3]

A comparison of shrimp consumption with consumption of sardines (representative of a rich catch and 1/5 the price of shrimps) showed that sardine consumption superseded that of shrimps in the minimum annual income bracket (within 2,650,000 yen) (Fig. 8.1). The other four income brackets with higher annual incomes consumed more shrimps than sardines.

In view of the foregoing facts, it is clear that domestic consumption of shrimps is higher in urban regions and among higher income households than in rural regions.

(3) Regionality of western high and eastern low

Figure 8.2 shows domestic consumption of shrimps according to regional blocks. Consumption is high in the western regions of Tokai, Kinki, Chugoku, and Shikoku and low in the eastern regions. It is difficult to determine the factors which influence regional variations in domestic

---

[3]Data from the Statistical Division of the Primer Minister's Secretariat and Statistical Report of the Ministry of Agriculture, Forestry and Fisheries, 1981.

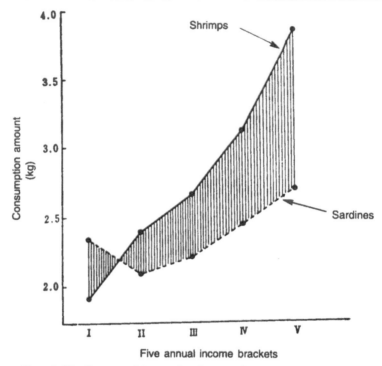

Notes: 1. The five annual income brackets are the same as those given in Table 8.4.
2. Consumption amount refers to annual consumption per household (1981).
3. Mean price for 100 g: shrimp 249.42 yen; sardine 41.46 yen.
Source: Statistical Division of Prime Minister's Secretariat (1981).
Fig. 8.1. Comparison of shrimp and sardine consumption according to income brackets.

consumption of shrimps. However, the higher consumption in western Japan might be attributable to better weather conditions and product availability.

Among the main shrimp-producing regions of Japan, Seto Inland Sea occupies first position, followed by the middle and southern regions of the Pacific Ocean. This trend continues even now. In 1980, of the total of 50,000 tons of shrimps produced in Japan, production in the Seto Inland Sea covered 42.2%, and the middle and southern Pacific regions 20.6%. Thus the Sea accounts for more than 60% of the total production.[4] In view of this fact, presumably the special liking for shrimps by the people in these regions, i.e., western Japan, has become traditionally stronger.

8.1.4 Future Trends in Consumption

The consumption of shrimps has grown steadily, supported by high-level

[4]Statistical Report on Fisheries Culture Production. Ministry of Agriculture, Forestry and Fisheries, 1980.

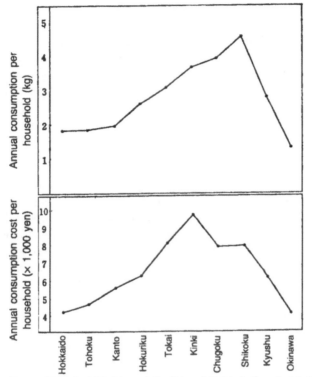

Source: Statistical Division of the Prime Minister's Secretariat, 1981.

Fig. 8.2. Regionality in shrimp consumption.

economic growth. The prime factor for this is the widening of effective demand for expensive fishery products which become affordable with an increase in personal income. As mentioned earlier, domestic shrimp consumption has a higher weightage in the urban high-income strata. However, the increase in shrimp consumption is also significant even among low-income groups and rural households. Hence there is no doubt that the overall upgrading of personal income forms the backdrop for increased shrimp consumption.

The demand for shrimps in major cities is mainly for business use. At the same time, improvement in personal income has popularized a life of leisure, which has encouraged eating outside the home, which in turn has widened the demand for business use.

Secondly, shrimps as food are highly acceptable to the Japanese people since the change in food prefere.ice and liking accord well with the taste of shrimps. From 1955 onwards, there has been a steady trend in the food habits of the Japanese toward westernization. The change from native food to western is particularly striking in young people. Shrimps are readily acceptable when boiled or fried. In this context, they differ from the common fishes, which are suitable only for native style cooking. Thus the consumability of shrimps has strengthened.

Thirdly, the food form of shrimps was acquired parallel to a decline in cooking efficiency within the household due to changes in the lifestyle of the people. In this context, it stands at par with the increasing demand for peeled tuna.[5]

The question now is whether the consumption trend hitherto will change and if so in what manner. This will depend on whether factors responsible for expanding consumption thus far will change and if so, will such a change continue.

As far as food preferences and ecological conditions are concerned, no radical change has been observed; in fact the status quo is expected to continue for some time. The question is an increase in national income, as this is greatly influenced by a favorable or unfavorable economy.

With the First Oil Crisis faced in November, 1973, the economic growth of Japan went into recession. Again, due to the Second Oil Crisis of 1979, economic conditions worsened. Under these circumstances consumer purchasing capability declined. This was reflected in the consumption rate of shrimps also. After peaking in 1975, consumption took a downward trend. It is assumed that the adverse economic conditions will continue for some years. The demand for high-grade fishery products cannot be expected to reach the level it did during the period of high economic growth. Therefore, the decline in shrimp consumption will also continue for some time. Concomitantly, a trend toward increased intake of animal protein has recently been observed. Farm products, the prices of which remain stable, are replacing fishery products. But at the same time concern for nutritional balance has attracted much attention, leading to renewed interest in fishery products. Thus the consumption of shrimps might not show a dramatic decrease.

8.2   CONDITIONS OF SHRIMP SUPPLY

8.2.1   Production within the Country and Quantity Imported

Along with increase in consumption of shrimps there was a steady increase year after year in supply. As shown in Table 8.5, the total supply before 1965 was less than 100,000 tons. Thereafter, it exceeded 100,000 tons, reaching 200,000 tons as of 1978. The export of shrimps was 1,000 to 4,000 tons, which, compared to the inland supply is negligible.

Comparing the quantity of domestic production with the imported quantity indicates that the former is gradually decreasing and the latter

---

[5]Details of this aspect are given on page 257 of the Proceedings of the Seventh Seminar held by the Fisheries Association, Tokyo University.

Table 8.5: Changing trends in domestic production, import and export quantities of shrimps (unit: 1,000 tons)

	Domestic production	Imported quantity (A)	Exported	Total supply (B)	(A)/(B) × 100	%age of shrimps vs total fisheries import
1961	75	4	1	78	5.1	30.7
1962	81	4	2	83	4.8	20.8
1963	88	12	1	99	12.1	39.5
1964	79	17	1	95	17.9	35.1
1965	68	21	2	87	24.1	34.6
1966	70	36	2	104	34.6	35.9
1967	62	44	1	105	41.9	41.6
1968	68	35	2	101	34.7	39.0
1969	59	49	3	105	46.7	46.7
1970	56	57	3	110	51.8	43.1
1971	51	79	4	126	62.7	47.8
1972	58	88	4	142	62.0	47.2
1973	63	117	4	176	66.5	35.8
1974	80	103	4	179	57.5	36.4
1975	70	114	3	181	63.0	35.7
1976	62	126	2	186	67.7	39.6
1977	54	128	2	180	71.1	33.4
1978	61	150	2	209	71.8	32.9
1979	54	164	2	216	75.9	33.6
1980	53	148	2	199	74.4	32.6
1981	56	162	3	220	75.9	31.8

Notes: 1. Domestic production includes that from aquaculture in shallow waters.
2. Imported and exported quantities are for fresh frozen shrimps.
3. Total supply equals total domestic production plus imports, minus exports.
Source: Domestic production provided by the Ministry of Agriculture, Forestry and Fisheries. The quantities of imports and exports were obtained from the Ministry of Finance.

steadily increasing. Between 1955 and 1965, domestic production was around 70,000-80,000 tons. Between 1965 and 1975, it decreased to 50,000-60,000 tons. This decline was apparently due to the pollution and destruction of coastal fishing grounds resulting from rapidly developing industries and extended development of coastal cities and towns, which were supported by high economic growth. Consequently production in the potential shrimp fishing belts of the Seto Inland Sea and Ise Bay had to be decreased or completely halted. It was around 1975 that culture measures based on the release of juvenile shrimps were actively developed. However, the resources have yet to make a complete recovery.

As against this condition, the amount of import was steadily increasing. Around 1970, it was 50,000 tons and soon increased to 100,000 tons and more. After 1978, it increased to 150,000-160,000 tons. Active import was initiated in October, 1961 when shrimps were freely imported together with other medium-grade fishery products. The prospects for imported commodities increased further concomitant with the ever-increasing demand, coupled with lack of domestic production. Major trading concerns and fishery products companies participated aggressively in the import business.

Thus in 1970, domestic production was subdued and the percentage of imports constituted more than half the total supply. Thereafter, it steadily increased and in recent years has exceeded 70%. In other words, imports are almost the core of fish supply.

Among the marine products imported into Japan, prawns and shrimps constitute the maximum total weight. During the first half of 1965 they constituted 40% of the total imported marine products. Thereafter they apparently declined relative to the increase in other imported items. Nonetheless, they constitute more than 30% and continue to dominate as the highest imported items.

8.2.2 Sources of Imports and Species Imported

The money involved in shrimp import was about 280 billion yen in 1971. The important source countries are arranged in descending order in Table 8.6. Except for Mexico, which holds ninth position, all the countries are in Asia and the South Pacific Ocean. First position is held by India, followed by Indonesia, Australia, and China. These four countries are very important import sources as they cover about 60% of the total import expenditure.

Table 8.6: Countrywise import (quantity and cost) of shrimps (fresh and frozen) (1981)

Position	Country	Quantity (ton)	Cost (1 million yen)	Percentage
	Total	167,129	279,642	100
1	India	40,368	54,005	19.3
2	Indonesia	24,311	44,066	15.8
3	Australia	13,309	31,284	11.2
4	China	14,954	30,251	10.8
5	Thailand	10,321	16,464	5.9
6	Taiwan	7,776	10,806	3.9
7	Malaysia	5,360	8,300	3.0
8	Pakistan	6,382	8,278	3.0
9	Mexico	3,173	6,989	2.5
10	Hong Kong	3,346	6,746	2.4

Source: Ministry of Finance.

Up to about 1965, China held first position; between 1965 and 1975 both China and Thailand lost much of their resources. Consequently, China's ranking dropped. On the other hand, Indonesia improved resource development and thus occupied a higher position. Later, in 1980, countrywise positions changed with Indonesia first followed by India,

China, and Australia. Subsequently shrimp resources began to decline in Indonesia. Moreover, the Indonesian government implemented strict regulations against shrimp trawlers, causing a decline in the import total. Indonesia's decline pushed India to first place.

Among imported prawns, headless unpeeled frozen products constitute the bulk. They are prepared by the packers of the place of origin or by joint enterprise. Recently, there has been a striking increase in the import of frozen peeled prawns from India.

The imported prawn species include *Penaeus indicus, Penaeus monodon, Metapenaeus dobsoni,* and *Metapenaeus affinis* from India. Those imported from Indonesia are mainly the white, pink, and black tiger prawns. From China, mostly frozen *Penaeus chinensis* is imported.

8.2.3 Importing Concerns and Importing Methods

Most of the concerns importing fishery products are members of the Cooperative of Marine Products Importers of Japan (established September 1, 1966). In April, 1983, the membership totaled 73 companies. Of these, 43 are in the shrimp import business. They include a composite concern and its branches (12), a major fishery products company and its branches (11), other trade companies (13), shrimp wholesalers (3), bulk vendors (2) and marine products companies (2). The import business of these concerns is very active and the money involved annually in import is 280 billion yen. Two other concerns dealing mainly in steel are also involved in the shrimp trade.[6]

In addition, several medium and small companies make spot purchases. These are not members of the Association. It is said that there are about 50 companies dealing in shrimp import. Of these, three are composite companies and two are major marine product companies.

The method of importing may be simple purchase or joint enterprise. Imports from India and China are of the first type and those from Indonesia mainly of the latter. Imports from other countries are a combination of the two.

Simple purchase import is a common method wherein a composite concern, major marine products company, their branches, and medium and small importers, as well as spot purchasers are directly involved in the import.

For example, most of the import from India is of the simple purchase type. Individual concerns enter into a purchase contract with the packer-cum-exporters of the concerned countries. With the advent of the season,

---

[6]Composite companies: Itochu, Tsumatsuesho, Sumitomo, Shoji, Tomen, Nisshoiwai, Nichimenjigyo, Marubeni, Mitsui Bussan, and Mitsubishi Shoji.
Major marine products companies: Kyokuyo, Taiheigyogyo, Nihon Suisan, Nihon Reizo.

the composite concern and the medium and small purchasers, totaling more than 30 companies, compete in the purchase at the site. Disputes sometimes arise as to the purchase price or observance of the contractual terms.

Import based on joint enterprise was initiated in 1955 by the major marine products companies. Composite companies also adopted this concept subsequently and this method of importation has gradually increased. As shown in Table 8.7, the number of joint enterprises has increased significantly from 1965. By June, 1979, the number had reached 40. By the end of 1979, the number of joint enterprises relating to fishery production was almost 220. Shrimp-related joint enterprises constituted 18%.

North America (USA and Canada) ranks first in terms of joint enterprises related to fishery products. With the acceptance of the 200-knot zone since 1977, in addition to systematic strengthening of fishing regulations in the North Pacific, joint enterprises have mushroomed for processing of salmon, trout, herring, and fish roe (herring, salmon).

It is obvious that joint enterprises set up for promotion of shrimp export have played a very important role in establishing such enterprises for other fishery products.

Regionally, shrimp-related joint enterprises are maximum in the Asian and Pacific regions, i.e., 27 enterprises constituting 68% of the total. Central and South America come next, followed by Africa. Thus most joint enterprises are located in developing countries but spread over 17 countries, with the maximum situated in Indonesia (12), followed by 6 in Papua New Guinea and 4 in Brazil. Only one or two enterprises have been established in the other countries.

As for the operations of the joint enterprises, these include fishing by shrimp trawler, freezing treatment and processing, and exporting to Japan. In other words, the operations include fishing, processing, and marketing, as exemplified by 29 enterprises, i.e., 73% of the total. Seven enterprises purchase the catch from operators at the fishing sites and after freezing and processing, export the product to other countries. Another 3 enterprises purchase processed products from the packers and export them. One American company has a joint enterprise for shrimp culture.

Details of the enterprise on the Japanese side of the joint venture are as follows: 26 major concerns, including composite concerns and their branches and major marine products companies and their branches, are the major investors. These companies constitute 65% of the total. Operations include fishing, processing, and export marketing. Medium and small marine product companies and traders make up the balance of 35%. They are mostly involved in processing and export or purchase-oriented sale.

Table 8.7: Number of enterprises involved in shrimp joint venture (June, 1979)

		Jointly by composite concern and major marine products company	Composite concern	Major marine products company	Others	Total
Regional	Asia, Pacific regions	12	3	3	9	27
	Central, South America	1		4	4	9
	Africa			3		3
	North America				1	1
	(Total)	13	3	10	14	40
Operational	Trawling, processing, sale	12	3	8	6	29
	Processing, sale	1		2	4	7
	Purchase-oriented sale				3	3
	Culture				1	1
Yearwise	Before 1959			3		3
	1960-1964	1		3	1	5
	1965-1969	2	2	1	1	6
	1970-1974	9		2	8	19
	After 1975	1	1	1	4	7

Note: The composite company and major marine products company include their branches.
Source: Suisansha (June, 1978).

When the capital outlay is large, the investment is mostly made in a cooperative manner by the main trading concern and its allies as well as the chief fishery concern and its allies. This trend rapidly progressed from the latter half of the 1960s. This type of cooperation between the trading concern and the fishery is effective and profitable in the management of the overall operations such as fishing, processing, and export-cum-sale. Thus for the trading concern the risk of unskilled fishing and processing aspects is precluded and the fishery concern enjoys the advantage of financial security, approval of acquisition of international rights and interests, and the establishment of a marketing system efficiently carried out by the trading concern.

8.2.4 Future Status of Imports

(1) Problems relating to shrimp resources in developing countries

Japan mainly imports shrimp from Asian and Pacific regions. As already mentioned, imports earlier depended on the developing countries of Central and South America and Africa.

The quantity of shrimp imported is presently 150,000 to 160,000 tons and has remained almost fixed for some years (see Table 8.5). This is because on the one hand the demand for shrimps has fairly well reached its limit and, on the other, resources of the countries from which shrimps are imported are considerably taxed.

World production of shrimps between the latter half of the decade beginning in 1970 and 1980 was 1,700,000-1,800,000 tons and appears to have been arrested at this level.[7] However, countries from which Japan mainly imports are already facing deterioration of resource conditions.

As mentioned above, the resources of China and Thailand which in 1965 ranked high, have gradually declined. From 1975, Indonesia and India have overtaken them. Even in these countries though the resource conditions are not absolutely encouraging. There is an overall downward trend. For example, in India the shrimp harvest in 1975 was 220,000 tons (maximum ever attained); thereafter the harvest began to decline and in 1978 was 180,000 tons, which further decreased in 1980 to 170,000 tons.[8]

Efforts are underway in these developing countries to develop culture industries as a means of increasing shrimp production. From the point of view of technology, apparently it will take quite some time to achieve en mass production of shrimps through the culture method. In India, aside from culture, efforts are also being made to increase production through deep-sea shrimp fishing.

In view of these facts, it may be concluded that as far as future shrimp resources are concerned, unless new areas in sea regions are explored in new countries, one can only expect a decreasing and certainly not an increasing trend.

The total supply of shrimps in Japan includes 75% imported. As there is no hope for a rapid increase in domestic production, the country cannot afford to reduce its dependence on imports for some years. Given these conditions, if the shrimp resources of the countries from which shrimps are imported decline, competition among purchasers will be heavy, leading to a tremendous increase in import price. There is also the apprehension of deterioration in quality. In fact, such adverse symptoms are already evident in countries such as India.

(2) Problems relating to economic policies in developing countries

In the developing countries the urge to develop unutilized marine resources along the coastal regions, expected to play an important role in the economic development of the country, have intensified with entry into the 200-knot era.

---

[7] FAO. Yearbook of Fisheries Statistics, 1980.
[8] The Marine Products Export Development Authority. Marine Products Export Review, 1981.

In the case of shrimp resources also, this same line of development has been encouraged. Along with shrimp fishing and developing the fishery processing industry, it is felt that the products exported to Japan will earn the foreign exchange needed for augmenting national economic progress. This is the main objective of government policy.

In Indonesia, this policy was implemented for procuring foreign exchange and to attain this objective, joint enterprises were established. The percentage of investment from the Japanese side for the joint enterprises was to be 49% within 15 years of the establishment (at the time of establishment, even 90%). It was emphasized that Japanese laborers and technicians would be replaced by Indonesians within 2-3 years, technical assistance matching the level of the country provided, and economic assistance given—all of which would form the basis for the economic development of the concerned country.

In India, the major products of the country are exported to obtain foreign exchange. With this foreign exchange, industrialization of the country can be planned. This forms the basis of the national policy. It is in this context that export items to Japan include shrimps after iron ore. Resource development, processing, and export have shown dramatic progress from the very outset. Joint enterprises were not encouraged. With promotion of coastal shrimp production and culture production by private fishing operators, development of freezing and processing industries for shrimps, and establishment of plants for checking the export shrimp products, progress in exports has been rapid. These several policies were recommended by the Marine Products Export Development Authority. A branch office of this authority is located in Tokyo, which ensures close interaction with the importers of Japan. Concomitantly, efforts are underway to increase exports by including marine products other than shrimp.

Thus there are considerable variations in the methods employed depending on the target country. In any case, the countries concerned aim to develop their economy by effective utilization of their resources and procurement of foreign exchange. On entering the 200-knot era, the fundamental level is expected to become more and more consolidated in terms of exports of marine products to Japan. Developing countries are well aware of this aspect.

It is necessary to have a perfect understanding of the economic policies of the developing countries; by supporting these policies, purchase-oriented import and joint enterprises developing import will show further progress. These are important conditions for sustainability of stable import of shrimps in Japan.

8.3   SCHEME FOR DISTRIBUTION OF SHRIMPS IN JAPAN

8.3.1   Distribution Circuit

Shrimps produced within the country are mostly dispatched fresh to the wholesale market in the respective native regions. Thereafter (1) they are transported to the wholesale market of the consumer region by the middleman; (2) diverted to the consumer region after processing by a qualified agency; and (3) put on sale either by small or large business groups after purchase by retailers and business traders. At the consumer wholesale market they are sold through middlemen to retailers and business organizations. In other words, the fresh shrimps produced within the country follow the same distribution circuit as other marine products produced indigenously. The only difference in the case of shrimps is a greater demand and consumption by businesses operating in large cities.

The shrimps produced within the country include *hokkaishima* shrimp of Hokkaido, *hokkokuaka* shrimp of the Hokuriku coast, *sakura* shrimp of the Tone estuary, and spiny lobsters of the middle, southern, and eastern coasts of the Pacific Ocean. These species pass through the wholesale markets of their respective native regions but most end up in a narrow distribution circuit for a particular business use and/or processing.

The distribution circuit of imported frozen shrimps which constitute 75% of the total supply, is shown in Fig. 8.3. The first step in their circulation route is passage from the importers (composite concern, major marine products company, other medium and small importers) to primary wholesalers on to central wholesale dealers, and then to frozen-food manufacturers. Thus they undergo a 3-stage sale.

The share of the primary wholesaler of imported frozen shrimps is large, constituting more than 70%, of the central wholesale market traders about 20%, while the remaining 10% goes to frozen-food manufacturers. The first stops for imported frozen shrimps are Tokyo and Osaka. Tokyo handles 60% and Osaka 40%. In the case of Tokyo, handling by primary wholesalers outside the shrimp market is strikingly high. In the case of Osaka, handling is mainly done by the central wholesale market dealers,

The primary wholesale vendors of out-of-market shrimps include the composite concerns and medium and small import companies covering 70-80%, direct import covering about 10%, and the central wholesale market operators covering 10-20%. Thus the composite concerns and the medium and small import companies are strikingly large in number. As against this, the vendors of the central wholesale market, wholesale operators, include composite concerns and medium and small import companies (50%), leading marine products companies (30%), and primary

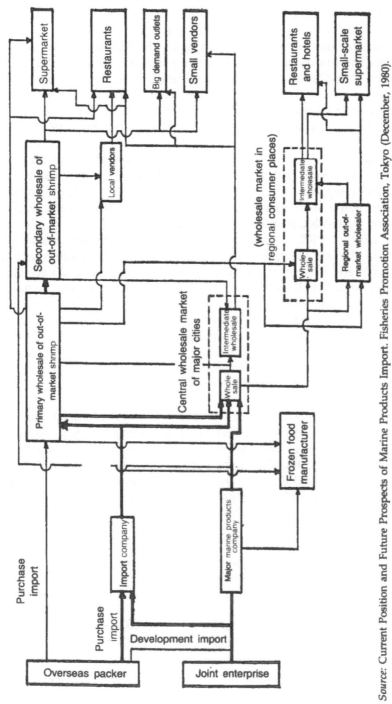

Source: Current Position and Future Prospects of Marine Products Import. Fisheries Promotion Association, Tokyo (December, 1980).

Fig. 8.3: Distribution circuit for imported frozen shrimps.

wholesale dealers of out-of-market shrimps (20%). Thus stocking by lead-ing marine products companies is significant.[9]

The manufacturer of frozen food is usually the shrimp-processing fac-tory under the control of the chief marine products concern or its serial companies. Therefore stocking is mostly done by the chief marine prod-ucts concern.

After passing through the first stage, circulation shifts to the second and third stages; on its way to terminal consumption the route divides into many branches.

There are about 20 small and large primary wholesale companies for out-of-market shrimps and their buyers include the secondary wholesale dealers (including local vendors), central and regional wholesale market operators, central wholesale market shrimp commission agents, restau-rants/hotel industries, and bulk-sale counters.

There are about 400 secondary wholesale concerns dealing in shrimps outside the market. The selling share of the primary wholesale concerns is about 30% and they act as the main route of circulation. At the second-ary wholesale outlets the consignments are classified according to quan-tity and size of the prevailing demand and then sold to small vendors, restaurants, hotels and bulk-sale outlets. In this case there is no self-storing and stocking is done in accordance with the orders received from hotels, etc. and then sold out. These are called local vendors. About 40 shops in Tokyo deal with 2,000 vendors selling *soba* (a noodle-like prepa-ration of buckwheat), tempura (oil fry), and *sushi* (special type of raw fish and shrimps).

It appears that the sale from the primary shrimp wholesalers out-of-market to the central and regional wholesale markets and wholesale deal-ers is about 10%. In the case of the central wholesale markets and whole-sale dealers of Tokyo and Osaka, the purchase is made directly from the importers and the purchase from the primary shrimp wholesaler outside the market is supplementary. In the case of the central wholesale market of medium and small cities and the wholesalers of the regional wholesale market, there is a higher degree of dependence on the primary shrimp wholesaler outside the market. The wholesalers of these markets sell directly to vendors, restaurants and bulk purchasers. In the case of cen-tral wholesale markets and wholesale dealers of Tokyo and Osaka, the sale is sometimes diverted to other central and regional wholesale mar-kets and wholesale dealers.

Regarding the central wholesale market professional middlemen op-erators, 25-30% is sold by the primary shrimp wholesaler out-of-market

---

[9]Report on Circulation Price of Frozen Marine Products. Circulation Economy Research Center, March, 1978.

and after passing through the secondary shrimp wholesale route, follows the main route. Particularly in the case of professional middlemen operators of the central wholesale market, purchases from the primary shrimp wholesaler outside the market (including the part purchased from the secondary shrimp wholesaler outside the market) is quantitatively larger than the purchase made from the wholesale dealer in the market. The route from the middleman operator to the final supplier is the same as mentioned earlier.

There is also the route of direct sale from the primary wholesale firms (outside the market) to restaurants/hotels and bulk-sale outlets. This is known as *teuri* in Japanese, meaning personal delivery, and provides delivery service. The targets are *soba* vendors, *sushi* vendors, and hotels, which constitute 20%. Bulk-sale outlets include department stores and supermarkets, constituting 10%.

The central wholesale market wholesale operation, a route of the first stage, sells frozen shrimps stocked from the importer (partly from the primary wholesale vendors of shrimps outside the market) to the local intermediate wholesalers. About 20-30% is sold to other market wholesalers, processing industries, and the secondary wholesalers of out-of-market shrimps. As for handling of the market, unlike the other fresh fishery products, circulation is from top to bottom at a relatively fixed price for sale and purchase.

The distribution/circulation routes of imported frozen shrimps within the country have thus been detailed. The most important feature is that circulation outside the market with primary wholesale of out-of-market shrimp at the apex, forms the main route, unlike other fishery products. The reasons for this type of wholesale circulation are given below.

Firstly, in the case of imported frozen shrimps, the species, quality, and size are prescribed and it is possible to negotiate price. Moreover, during the course of continuous stabilized import, the brand of the packer gets established and deals settled on the basis of the brand. Hence there is no need to collect the consignments at a specified place, examine and decide the product, and fix the price, as done in the case of so-called wholesale market circulation.

Secondly, even after 1965 when imports of frozen shrimp positively increased, the conventional style of fresh shrimp dealing was pursued in the central wholesale market. It was not a pragmatic approach. Therefore, in keeping with the increased supply of imported frozen shrimp, circulation outside the market accelerated.

Thirdly, the importer with the composite firm as the main body emphasized establishment of the internal sale route independent of the internal sale of the imported frozen shrimp. It is in this context that the professional wholesale of shrimp outside the market was positively cultivated.

Finally, there is an endless demand for imported frozen shrimp for business use, mainly for hotels and restaurants. To meet such demands the wholesale route with delivery service is excellent.

8.3.2   Trend of Primary Wholesale of Shrimp Outside the Market

The most important feature of the internal circulation of imported frozen shrimp is that it has the wholesale route with primary wholesale outside the market at the apex.

There are just over 20 primary wholesalers outside the market. These include 11 leading professional shrimp wholesalers with dealings in imported frozen shrimp exceeding 10 billion yen per annum. The total sum involved in the dealings by primary wholesalers exceeds 200 billion yen and the amount of dealing covers 65-70% of the total amount of imported frozen shrimp.[10]

Of the 11 companies, 4 are importers which are offshoots of the leading composite trading company. They are developing the internal sale of mainly the imported frozen shrimp of a particular leading composite trading company.

The other 7 companies are independent shrimp wholesalers. They stock imported frozen shrimp from various importing companies. One or two even have a strong link with a particular composite trading company.

Of the 11 companies, 3 (one offshoot of a leading composite trading company and two independent wholesalers) do not restrict themselves to internal sales, but also carry out import business.

The predecessors of most of the big independent wholesale dealers of shrimp were middlemen professionals in the central wholesale market, chiefly involved in selling the fresh shrimps produced in Japan to hotel and restaurant industries.

After World War II the production of fleshy prawns (*P. chinensis*) was very popular in western waters between 1950 and 1960. During this period they were even exported to America. However, once Japan entered the free import era after 1961, emphasis shifted to handling imported frozen shrimps through the out-of-market route. The net sale was established and professional wholesalers were cultivated.

Recently, the quantity of imported frozen shrimps has stagnated at around 150,000 tons. On the other hand, along with the prolonged period of unfavorable conditions, the demand by restaurants and hotels became a headache. Under these circumstances a new movement developed in

---

[10]Leading professional shrimp wholesalers: (1) Independent—Daiedaigen, Kairo no Daimaru, Kairojo, Kairosei, Tokyo Kairo Shoji, Chusho, Daito; (2) companies under the leading composite trading company—Taiyonosuisan (Mitsui Bussan), Kosui (Mitsubishi Shoji), C.I. Sea Foods (Itochu), Kairoko (Marubeni). All are limited companies.

the internal circulation mechanism of shrimps. One step in this regard was to establish an independent internal sale mechanism by the composite trade companies with expertise for import and leading marine product companies. Already four of the leading shrimp wholesalers are the offshoots of a leading composite trade company. Other new composite trade companies and leading marine product companies are participating in the primary shrimp business outside the market.

Following this trend, it is hoped that there will be a strong positive association among the independent leading shrimp wholesalers and other medium and small wholesalers of shrimp outside the market with a series of importers, such as composite trade companies or leading fishery product companies. At the same time, by increasing the handling quantity of frozen products other than shrimps, composite wholesalers of frozen fishes will be able to effectively maintain their independence. The internal sale of shrimps could thereby once again gradually show an upward trend.

8.3.3 Important Points Relating to Internal Distribution

Of the 16 items of fresh fishes and shellfishes surveyed by the Statistical Division of the Prime Minister's Secretariat (1981), domestic consumption of shrimp was found to occupy 5th place. Again, according to a survey carried out in Toyohashi city with a population of 300,000, shrimp consumption outside the home occupied 3rd place after tuna and squid.[11]

Thus, in the food habits of the Japanese people, since shrimp is a high-grade fishery product it is very popular and the demand for it always constant.

A stable supply of shrimps to the consumers is dependent, however, upon a stable supply of fishery products.

Certain precautions in ensuring a stable supply of shrimp must be taken, primary among them being quality. Shrimps are distributed either fresh or frozen. Hence the degree of freshness is very important. Imported products from tropical developing countries are quantitatively large but often experience problems with maintenance of hygiene. Therefore a strict quality check is mandatory. Moreover, this is not confined just to freezing maintenance and preservation during the course of internal circulation. Quality should be controlled right from the production point in the country of shrimp origin.

For Japan, India is the country from which the maximum quantity of shrimp has been imported over the years. Strict regulations were imposed in India at the governmental level during the 1960s. In September, 1963, the Voluntary Preshipment Inspection became mandatory and from

---

[11]Report of the Food Demand and Supply Research Center (1980).

January, 1964, the export inspection office was established. In this way, not only frozen and canned shrimps were inspected before exporting, but also all the major export fishery products likewise carefully examined for quality. In December, 1979, a new regulation called *quality control* during the course of production was introduced for frozen shrimps, lobsters, butterfish, and Mongolian squid. This regulation ensures implementation of quality control at every stage from fishing to ship loading.

In other countries also where shrimps are produced and exported, quality-control measures have considerably advanced. These measures are expected to be finalized very shortly.

The second precaution concerns the supply of items suitable for use. Fresh frozen shrimps are utilized in various ways, both for domestic consumption and for business purposes. Therefore, it is necessary to supply them in such a manner that all aspects of use will be satisfactorily met. Particularly in the case of business use, the species and size differ according to whether they will be used by the *soba* vendor, the *sushi* vendor, high-grade food outlets, fast-food joints, etc. The species will also differ depending on the season and even on a day-to-day basis. Thus it is imperative that the supply system carefully take into consideration a diverse consumption pattern. The role of direct delivery and local sale counter centers based on professional wholesalers of shrimps outside the market is extremely important.

The third point of concern pertains to a stable supply price. During the high-grade economic growth period in Japan, shrimp consumption, in spite of the high price, increased because the national income was high and the out-of-home food industry developed rapidly. However, as the economy entered recession, domestic consumption almost ceased and business consumption as well as the out-of-home food industry became highly unstable and limited. High prices curtailed buying. So even given their strong desire for shrimp, consumers will avoid them if the price is out of reach and switch over to consumption of cheaper fishery products as well as meat products. Furthermore, an upswing in prices causes considerable distress in the fishing industries.

As for domestic circulation, insofar as possible, speculative dealings should be avoided and concomitantly complicated branch circulation routes should be simplified.

Frozen shrimps can be preserved and therefore easily handled in speculative trade. However, the circulation industry feels that conditions are not yet congenial for handling frozen shrimps on a speculative basis.

As mentioned above, the importers, that is, the composite trade companies and leading marine product companies, tend to establish a simplified internal sale system which could otherwise become cumbersome. But since consumer use of shrimps is diverse the wholesale function has necessarily to meet a complicated demand. Only through mutual cooperation among wholesale dealers of shrimps outside the market, can severe competition be avoided and insofar as possible the circulation route kept simple.

Feed for Prawn Culture

9.1 NUTRIENT REQUIREMENTS OF PRAWNS

9.1.1 Nutrient Requirements of Cultured Prawns

(1) Protein

Protein is an essential nutrient for growth and sustenance of life. It plays an important role as a building component of the body. Deshimaru (1981) used casein and egg albumin (9 : 1) as protein and observed the growth rate and feed efficiency in *Penaeus japonicus*. He concluded from the results that the optimum level of feed protein is 52-57%. The optimum quantities of protein to be added in the diet of various species of prawns are given in Table 9.1. It can be seen that the optimum protein content in prawn feed varies widely (30-57%) depending on the species. One factor in this variation is mode of feeding—carnivorous, herbivorous, or omnivorous. The protein requirement of prawns also varies according to the type of protein and the contents of essential amino acids and their

Table 9.1: Protein requirements of prawns

Types	Optimum protein content (%)	References
Penaeus japonicus	52-57	Deshimaru and Yone (1978)
P. indicus	43	Colvin (1976)
P. monodon	46	Lee (1971)
	40	AQUACOP (1977)
	40	Khannapa (1977)
	35	Bages and Sloane (1981)
P. aztecus	23-31	Shewart et al. (1973)
	40	Venkataramiah et al. (1975)
P. setiferus	28-32	Andrews et al. (1972)
P. californiensis	31	Colvin and Brand (1977)
P. stylirostris	35	Colvin and Brand (1977)
P. vannamei	30	Colvin and Brand (1977)
P. merguiensis	50	AQUACOP (1978)
	34-42	Sedgwick (1979)
Metapenaeus monoceros	55	Kanazawa et al. (1981)

percentages. In the case of *Penaeus duorarum* even soybean protein serves as an effective source of protein in the feed. As a protein source for lobsters (*Homarus americanus*), casein and soybean protein proved better than fish meal and shrimp meal added to casein and soybean protein significantly improved growth and survival percentages. It was further reported that crab meal is the best protein source and should be used as the standard test protein for crustaceans.

(2) Essential amino acids

To determine the amino acid requirement of *Penaeus japonicus*, an experimental amino acid diet was prepared by replacing the protein in the refined diet with an amino acid mixture and the prawns reared on this diet. Prawn growth was extremely poor and mortality very high. Acetate-^{3}H was administered to the prawns and the intake of amino acids constituting the body protein observed.

It was concluded from the results that a total of 10 amino acids, namely, arginine, methionine, valine, threonine, isoleucine, leucine, lysine, histidine, phenylalanine, and tryptophan are essential. It was further found that the same 10 amino acids are essential for *Astacus astacus, Palaemon serratus, Penaeus aztecus, Macrobrachium chione,* and *Macrobrachium rosenbergii.*

(3) Sterol

Animals are usually capable of synthesizing sterol in the body from acetate. However, in all the crustaceans such as crabs, blue crabs, spiny lobster (*Panulirus japonicus*), and prawns examined for the induction from acetate-^{14}C, that is, meparonate-^{14}C into cholesterol, it was observed that they lack the capacity to synthesize sterol (Table 9.2). For the growth of crustaceans, cholesterol appears to be an essential nutrient. In fact, culture experiments confirmed the requirement for sterol for lobsters, *kuruma* prawns, and *Metapenaeus monoceros*. The optimum cholesterol content in the diet of lobsters, *Penaeus japonicus,* and *Metapenaeus monoceros* is 0.5%.

To ascertain the utility of sterols other than cholesterol, refined diets supplemented with ergosterol of yeast or stigmasterol of plants or β-setosterol were prepared and the prawns (*Penaeus japonicus*) reared on them. Though all the three sterols were utilized, growth was inferior compared to a cholesterol-supplemented diet. In the case of crustaceans, be it ergosterol or 24-methylene cholesterol of C_{28} or β-setosterol of C_{29}, all are utilized only after conversion into cholesterol of C_{27} in the body. This was confirmed through tracer experiments using radioisotopes. The conversion of sterols of C_{28} and C_{29} into C_{27}-sterol is by the methyl removal at the C_{24} position. Crustaceans differ from other organisms in possessing the conversion enzymes for this purpose. In higher animals, cholesterol is an important substance as a precursor of bile acid and steroid hormones. Experiments on spiny lobsters have revealed that even in the case of crustaceans, cholesterol serves as a precursor for sex

Table 9.2: Sterol biosynthesis of crustaceans

Types	Capability for biosynthesis	References
Astacus astacus	—	Zandee (1962, 1964, 1966)
Cancer pagrus	—	Oord (1964)
Homarus gammarus	—	Zandee (1964, 1967)
Astacus fluviatilis	—	Gosselin (1965)
Rhithropanopus herrisii	—	
Libinia emerginata	—	Whitney (1969)
Callinectes sapidus	—	
Balanus mubilus	—	Whitney (1970)
Artemia salina	—	
Penaeus japonicus	—	
Panulirus japonicus	—	Teshima and Kanazawa (1971)
Portunus trituberculatus	—	
Carcinus maenas	—	
Eupagurus bernhardus	—	Walton and Pennock (1972)
Armadilidium unlgare	—	
Cirolana halfordi	—	O'Rourke and Monreo (1976)
Ligia occidentalis	—	
Helice tridens tridens	—	
Sesarma dehaani	—	Teshima et al. (1976)

hormones and the adrenocortical hormone. Crustaceans grow by periodic ecdyses and the ecdyson which initiates the ecdysis is biosynthesized from cholesterol—a fact confirmed in the study of lobsters.

(4) Essential lipids

The presence of essential lipids in crustaceans, as in the case of fishes, has been confirmed through the study of lipid biosynthesis and also through rearing experiments.

When acetate-^{14}C or palmitic acid-^{14}C was administered to *Penaeus japonicus*, *Homarus gammarus*, *Penaeus monodon*, and *Penaeus merguiensis*, it was readily inducted into saturated fatty acids and monoenic acid but there was hardly any induction into high-grade unsaturated fatty acids such as linoleic (18: 2n-6), linolenic (18: 3n-3), eicosapentanoic (20: 5n-3), and docosahexanoic acid (22: 6n-3). The results appear to indicate the possibility of these fatty acids being essential fatty acids for crustaceans.

On the basis of rearing experiments, it was confirmed that prawns and their allies require high-grade unsaturated fatty acids as essential lipids. When linolenic acid was added to the refined synthetic diet of *Penaeus aztecus*, weight gain improved. The requirement is 1-2%. For *Penaeus japonicus* also linoleic acid, linolenic acid, eicosapentanoic acid, and docosahexanoic acid are essential lipids. n-3 linolenic acid is more effective as an essential fatty acid than n-6 linoleic acid. The optimum content of these acids in the diet of the prawns is around 1% (Fig. 9.1).

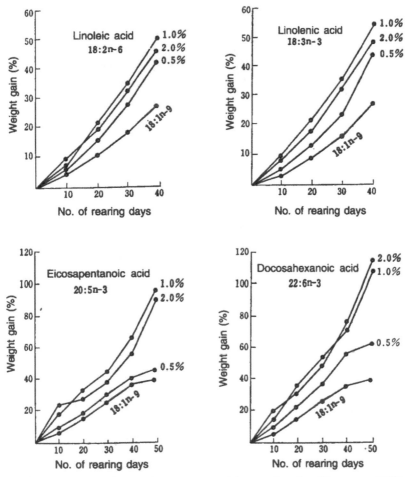

Fig. 9.1. Essential fatty acid requirement of *Penaeus japonicus* (Kanazawa, 1979).

(5) Phospholipid

In the case of juvenile prawns (*Penaeus japonicus*), the fat extracted from clams has an excellent growth accelerating effect compared to Alaska pollack-liver oil. When the clam oil was analyzed and the effects of the various lipids on the juvenile prawns (0.4-1.0 g body weight) compared, it was found that the group reared on the diet supplemented with 7% Alaska pollack-liver oil + 1% lecithin fraction had better weight gain than the group reared on the diet supplemented only with 8% Alaska pollack-liver oil + 1% cephalin fraction. Thus clam oil has excellent nutrient value for *Penaeus japonicus*. This is because clam oil is rich in n-3 high-grade unsaturated fatty acids. It likewise appears that the action of phospholipid itself plays an important role in this context.

When juvenile prawns of *Penaeus japonicus* (1 g) were reared for 30 days on a diet without phospholipid and a diet with 3% soybean lecithin, it was found that weight gain decreased strikingly in the group reared on a diet without phospholipid (Fig. 9.2). Moreover, compared to the group reared on a diet with phospholipid, the group denied it showed low diet efficiency and survival percentage (Table 9.3).

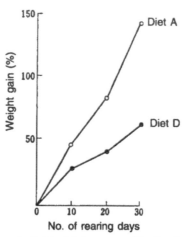

Diet A: phospholipid added;   Diet D: phospholipid lacking

Fig.9.2. Effect of phospholipid on growth of *Penaeus japonicus* (Teshima, 1985).

Table 9.3: Results of rearing *Penaeus japonicus* using diet supplemented with phospholipid (Diet A) and without it (diet D) (Teshima, 1985)

Items	Diet A	Diet D
Penaeus japonicus (no. of prawns)	15	15
Duration (days) of rearing	30	30
Average weight (g): initial	1.03	1.06
final	2.51	1.72
Weight gain (%)	144	62.3
Survival (%)	100	80
Diet efficiency[1]	0.49	0.22
Accumulation of diet fat (%)[2]		
Total fat	41.8	30.3
Phospholipid	57.0	—
Cholesterol	88.0	32.3
Neutral fat[3]	29.0	21.6

[1] Weight gain (g)/quantity of ingested feed (g).
[2] Accumulation (%) = A/B − C × 100.
      A: Intake of fat component of diet/one specimen (*Penaeus japonicus*).
      B: Fat component of prawn tissue at commencement of experiment/one specimen (*P. japonicus*).
      C: Fat component of prawn tissues at end of experiment/one specimen (*P. japonicus*).
[3] Neutral fat other than cholesterol.

Conklin *et al.* (1980) reported that the survival percentage of juvenile lobsters reared on a diet with 7.5% soybean lecithin increased markedly (Table 9.4). Phosphatidylcholin is essential for the survival of lobsters and cannot be replaced by products of alkali hydrolysis, lipids or other phospholipids. At the same time, lecithin containing high-grade unsaturated fatty acids is said to be most effective. It was recently reported that when crab meat is used as the protein source for lobsters there is no need for phospholipid.

Table 9.4: Effect of adding soybean lecithin to diet on survival rate of lobsters (Conklin *et al.*, 1980)

Diet	Survival (%)		
	30 days	60 days	90 days
5 % soybean lecithin	46	0	—
7.5 % soybean lecithin	100	97	89
10 % soybean lecithin	100	86	*
20 % soybean lecithin	97	86	*
35 % soybean lecithin	100	89	*
Soybean lecithin not added	56	0	—

*Rearing discontinued after 60 days.

(6) Carbohydrates

As for the carbohydrate nutrient requirement of prawns, *Palaemon serratus* is said to readily digest wheat starch, dextrin, and glycogen. *Penaeus aztecus* grows well on a diet with low protein and high carbohydrate. However, growth is inhibited when the diet is supplemented with 20% glucose. *Penaeus duorarum* grows better on a 10% glucose diet than on a 10% starch diet but growth is inhibited when 40% glucose is added to the diet. Growth of *Penaeus japonicus* is also inhibited on a 20% glucose diet. In the case of *Penaeus japonicus*, disaccharides such as cane sugar, maltose, and trehalose, and polysaccharides such as starch, dextrin, and glycogen are a better carbohydrate source than monosaccharides such as glucose, galactose, and fructose (Fig. 9.3). For improved growth of *Penaeus monodon*, cane sugar and dextrin are said to be good. Inhibition of prawn growth with the addition of glucose but not with the addition of disaccharides or polysaccharides is explained as follows: dietary glucose is rapidly absorbed through the digestive system and released into the blood but this high concentration of glucose is not well utilized in energy metabolism; disaccharides and polysaccharides on the other hand, after digestion, are absorbed in installments in the form of glucose from the digestive canal. Regarding the diet of *Penaeus japonicus*, the optimum contents of carbohydrate and protein are maltose 24% and protein 40%. Protein is found to have a restricting action on maltose.

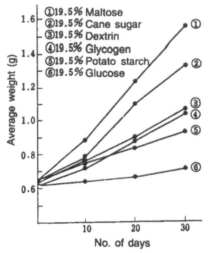

Fig. 9.3. Effect of various carbohydrates on the growth of *Penaeus japonicus* (Abdel Rahman *et al.*, 1979).

The body of a prawn is enveloped in a chitinous exoskeleton. To enable the prawn to grow, this exoskeleton is periodically sloughed in a phenomenon called ecdysis. Chitin, the chief component of the exoskeleton, is synthesized from glucose through glucosamine. It has been reported that when chitin is added to the diet growth deteriorates, but the positive effect on *Penaeus japonicus* with the addition of glucosamine has been recognized.

(7) Minerals

It has been reported that at the time of ecdysis naturally grown lobsters need supplementary feed rich in calcium content. A high calcium diet does not affect the growth rate and survival percentage of the lobsters but the mineral metabolism of the exoskeleton is accelerated. Castell *et al.* have reported that there is hardly any effect even when the synthetic diet of lobsters lacked minerals. The optimum ratio of calcium-phosphorus for the growth and survival of lobsters was 1 : 2 (Table 9.5). Calcium for *Penaeus aztecus* is supplemented from sea water. However, since sea water contains very little phosphorus, it has to be added in the diet. The fact that *Penaeus japonicus* absorbs calcium from sea water has been confirmed through radioisotope experiments. The optimum quantities of minerals to be added in the diet for *P. japonicus* are 1% calcium, 1% phosphorus, 0.3% magnesium, 0.9% potassium, and 6 mg% copper. The ideal ratio of calcium and phosphorus is 1 : 1. Between the calcium supplemented group and the calcium-denied group no significant difference was observed. However, in order to obtain a 1 : 1 ratio of calcium: phosphorus, the

Table 9.5: Influence of calcium-phosphorus ratio in the diet on growth and survival of lobsters* (Gallagher *et al.*, 1978)

Calcium-phosphorus ratio	0.37	0.51	1.01	1.55	1.81	1.91
Body length (mm)	10.8	11.4	11.1	9.0	10.1	10.0
Weight (g)	0.86	0.90	0.94	0.46	0.70	0.64
No. of ecdyses	4.9	5.2	5.2	4.9	4.1	4.2
Survival (%)	61	89	64	20	39	28

*90 days of rearing.

quantity of calcium to be added is 1%. In the case of fishes, the utility percent differs among first-grade phosphate, second-grade phosphate, and third-grade phosphate, which greatly influence growth of flesh and skeleton. It is necessary to study the effect of various inorganic phosphates on prawns also. In the case of copper, there is almost no difference in growth rate between a diet with 6 mg% copper and a copper-free diet, but growth is inhibited if the copper concentration is increased above 6 mg%. The addition of iron and manganese also inhibits growth of *P. japonicus*.

(8) Vitamins

Conklin *et al.* (1980) studied the influence of vitamin mixture on the growth rate and survival of lobsters and reported that the optimum quantity to be added is 4%. It was recently reported that lobsters are capable of biosynthesizing vitamin C and do not require its addition to their diet. *Penaeus japonicus* requires 0.06% choline and 0.2% inositol in its diet, as well as 6-12 mg% thiamine and 12 mg% pyridoxine. Spiny lobsters and *kuruma* prawns almost lack biosynthesis of vitamin C and so it has to be added in their diet. The ideal quantity of vitamin C to be added in the diet of *Penaeus japonicus* is about 1% taking into consideration weight gain, survival percentage, and vitamin C concentration in the hepatopancreas. Vitamin C is unstable both during preparation of the diet and during subsequent preservation. Specifically, during preparation 40% of the quantity added gets destroyed due to heat treatment. Hence the actual ingestible quantity of vitamin C for *P. japonicus* is considered to be less than 0.5%. Vitamin C deficiency can be prevented by applying 0.022-0.043% of the highly stable vitamin C, L-ascorbic acid-1-phosphoric acid-magnesium. A vitamin C deficiency in *P. japonicus* is readily recognized when the lateral and abdominal muscles (flesh) are pale gray. The so-called "black deaths" of *Penaeus californiensis* and *Penaeus stylirostris* wherein the cuticular layer, esophageal wall, stomach wall, hindgut wall, and gills turn black is, according to Lightner *et al.*, due to vitamin C deficiency.

9.1.2   Nutrient Requirements of Larval Prawns

After hatching, larvae of prawns undergo repeated metamorphoses in a relatively short period when they pass through nauplius, zoea, and mysis stages. Their nutrient requirements have obviously to be different from those of juvenile forms. Unfortunately, studies of this aspect are scant. But Kanazawa *et al.* (1980-1985) developed an artificial diet to replace the biological diet of crustacean larvae, enabling detailed research on their nutrient requirements.

(1) Protein

Teshima *et al.* (1981-1985) studied the effect of various proteins and crystalline amino acids using a fine granular artificial diet. As protein source for the larvae of *Penaeus japonicus*, casein, casein-gelatin (3 : 1), and white fish meal have a higher nutritional value than gelatin, albumin, and an amino acid mixture. The amino acid composition of body protein of zoea II and casein were compared and arginine, considered to be extremely insufficient in a casein diet, was added in the form of crystals. The supplementary effect was confirmed. In the case of larvae, the nutritional value of casein is improved by adding arginine chlorate crystals and growth as well as survival were the same as for the control group (fed with *Chaetoceros*).

   To obtain a basic idea about the use of fat and carbohydrate as the energy source and the optimum protein content of the diet used for *Penaeus* larvae, the protein, fat (lipid), and carbohydrate contents in the callagenan granular diet, respectively coded as P, L, and C, and the effect of these factors on growth and survival of *Penaeus* larvae were studied. Rearing experiments were conducted in two phases. The protein, fat, and carbohydrate contents were adjusted by using casein, cod-liver oil, and glucose—cane sugar—alpha starch in the ratio of 5.5 : 10 : 4. When the carbohydrate content was fixed at 15%, and the protein content gradually increased from 25 to 35 to 45 and 55%, the feed effect gradually increased and the survival percentage was 55%. The growth index attained the maximum level of 45% (Fig. 9.4). On the other hand, no significant conclusion could be drawn regarding the lipid factor and the mutual action of P × L. When the diet contained 15% or more of digestible carbohydrates, even by adding more than 6.5% lipid, the feed effect did not increase. In the case of a diet with lipid content fixed at 6.5%, the survival percentage of the larvae of *Penaeus japonicus* changed significantly according to the protein content (35, 45, 55%) and carbohydrate content (5, 15, 25%). In fact a significant effect was also observed in the mutual action of P × C. When the carbohydrate content was 5% and the protein content of the diet increased, a progressive improvement in larval survival was observed. When the carbohydrate content was changed to 25%, 15%,

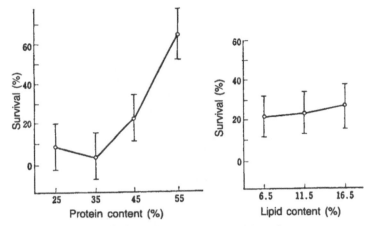

The digestible carbohydrate content, comprising glucose—cane sugar—alpha starch (5.5:10:4) was fixed at 15%. Protein and lipid content (casein PLO) were changed. All the diets contained n-3-HUFA (0.5%) and soybean phospholipid (2%) as lipid source other than PLO. I—○—I: central (95%).

Fig. 9.4. Influence of protein and lipid content in the diet of *Penaeus* larvae on survival percentage (Teshima and Kanazawa, 1984).

and 5%, it was found that the protein content should be about 45%, 45-55%, and 55% respectively. Between 35% and 55% no significant change in survival percentage was observed. In view of these results, it is clear that when the lipid content is 6.5%, the optimum protein content for the larvae varies according to the carbohydrate content and when the carbohydrate levels are 25%, 15%, and 5%, the corresponding protein contents should be about 45%, 45-55%, and 55% respectively.

(2) Fat (Lipid)

It is well known that crustaceans lack the ability to synthesize sterols. Therefore, for the purpose of growth they require sterols in the feed. For the larvae of *Penaeus*, sterols are essential nutritional elements. According to rearing experiments using the callagenan granular diet, the growth of larvae fed on a sterol-free diet is inferior to that of larvae fed with a cholesterol-supplemented diet. This is also true of the survival percentage. The optimum cholesterol content in the diet is about 1%. When various types of C_{27} - C_{29} sterols were used as the sterol source to observe their comparative effect, it was found that the nutritional value of all the sterols was inferior to cholesterol. 24-methylene cholesterol, ergosterol, 24-methylcholester-5,22-dienol, and isofucosterol have a markedly high feed effect (Fig. 9.5). Thus it was concluded that the larvae of *Penaeus japonicus* convert the external C_{28}, C_{29}-sterols to cholesterol inside the body. The zoea of *P. japonicus* are incapable of deriving linoleic acid and linolenic acid from palmatic acid (16:0) in the feed ($1-^{14}C$). For the pur-

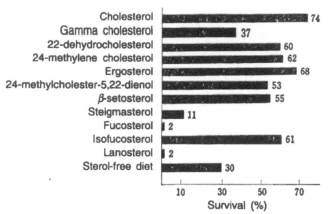

Fig. 9.5. Effect of sterols on the survival of *Penaeus* larvae (Teshima, 1985).

pose of growth, it appears they require n-3 high-grade unsaturated fatty acid. Actually, the essential fatty acids for *Penaeus* larvae have been confirmed to have an effect in the following order: eicosapentanoic acid $\rightleftharpoons$ docosahexanoic acid > linoleic acid > linolenic acid. The optimum concentration in the diet is said to be 1% (Fig. 9.6).

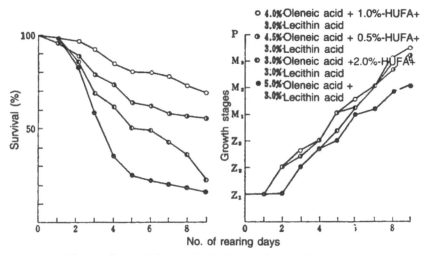

Fig. 9.6. Essential fatty acids for *Penaeus* larvae (Kanazawa, 1981).

For the growth and survival of *Penaeus* larvae also, the importance of phospholipid is confirmed. It is evident from Table 9.6 that even when Alaska pollack-liver oil was added to the diet, in the absence of phospholipid all the larvae died at mysis-1 stage. Regarding the effect of various phospholipids on *Penaeus* larvae, the addition of 1% soybean phosphatidylcholin, skipjack egg phosphatidylcholin, and soybean

Table 9.6: Effect of fatty acids on growth and survival of *Penaeus* larvae (Kanazawa, 1983)

Lipid in diet	Growth index*	Survival (%)	No. of observation days**
7% PLO + 1% chicken egg PC	4.5	0	6
7% PLO + 1% skipjack egg PC	6.4	91	6
7% PLO + 1% soybean PC	6.0	72	6
7% PLO + 1% sheep brain PE	5.3	34	6
7% PLO + 1% skipjack egg PE	4.7	48	6
7% PLO + 1% sheep brain SM	4.0	0	6
7% PLO + 1% sheep brain PS	5.7	55	6
7% PLO + 1% soybean PI	7.0	88	6
8% PLO	4.5	7	6

PLO: Alaska pollack-liver oil; PE: phosphatidylethanolamine; PS: phosphatidylselin; PC: phosphatidylcholin; SM: sphengomielin; PI: phosphatidylinositol
*Growth index: 1, zoea 1; 2, zoea 2; 3, zoea 3; 4, mysis 1; 5, mysis 2; 6, mysis 3; 7, postlarvae 1.
**No. of observation days: in the experimental group the duration covered zoea 1 to postlarvae 1

phosphatidylinositol proved effective. However, chicken egg phosphatidylcholin, sheep brain phosphatidylethanolamine, skipjack egg phosphatidylethanolamine, sheep brain phosphatidylgeline, and sheep brain sphengomielin proved ineffective for either growth or survival. The following conclusions can be drawn from these results:

a) The presence of cholin or inositol in the phospholipid molecules is effective.

b) The presence of high-grade unsaturated fatty acids, linoleic acid and linolenic acid, eicosapentanoic acid, and docosahexanoic acid in the phospholipid molecule is effective.

c) The requirement of phospholipid is about 0.5-1.0%. This is because *Penaeus* larvae appear to require phospholipid as an essential nutritional component.

   i) In the case of crustaceans, the fats move (migrate) in the form of phospholipid.

   ii) For cholesterol metabolism, phospholipid containing unsaturated fatty acids is used.

   iii) The enzymatic activity for phospholipid synthesis is extremely weak in the zoea and mysis stages.

(3) Carbohydrates

As for the effect of carbohydrates on the growth of *Penaeus* larvae, it was found that starch and dextrin are best, followed by cane sugar. Monosaccharides such as glucose, etc., are not effective for growth and, in fact, growth is mostly arrested at the mysis-3 stage.

(4) Vitamins

As for the requirement of vitamins for the growth of *Penaeus* larvae, it was found that a deficiency in any one of the following would result in high mortality: alpha-tocopherol, nicotinic acid, choline, pyridoxine, biotin, folic acid, ascorbic acid, cyanocobalamin acid, calcipheral, inositol, menadione, riboflavin, thiamine, and β-carotene. Growth is arrested at the mysis-2 or mysis-3 stage. In any case, growth never attains the postlarval stage (Table 9.7).

Table 9.7: Vitamin requirement of *Penaeus* larvae

Vitamins not included [1]	Survival (%) [3]	Vitamins not included	Survival (%)
Control group[2]	92	Ascorbic acid-Na	8
Calcium pantothenate-Ca	89	Cyanocobalamin acid	7
Alpha-tocopherol	49	Calcipheral acid	5
Nicotinic acid	25	Inositol	0
Choline chloride	15	Menadione	0
Pyridoxine acid-HCl	15	Riboflavin	0
Biotin	12	Thiamine-HCl	0
Folic acid	8	β-carotene	0

[1] Vitamin removed from the control group (complete diet with all the vitamins).
[2] Control group: The following vitamins are present in callagenan MBD (D-2) (mg/100 g diet: total 3.17 mg/100 g): ρ - amino aromatic acid, 10; biotin, 0.2; inositol, 200; nicotinic acid, 40; pantothenic acid Ca, 60; pyridoxine-HCl, 12; riboflavin, 8; thiamine-HCl, 4; cyanocobalamin, 0.08; ascorbic acid-NA, 1,000; folic acid, 0.8; choline chloride, 600; menadione, 4; β-carotene, 9.6; α-tocopherol, 20; calcipheral, 1.2.
[3] Survival (%) taken after 9 days (larvae of control group attain postlarval stage).

9.2   REARING DIET OF PRAWNS

9.2.1   Early Period Biodiet of Prawns

During the zoea and mysis stages of *Penaeus* larvae, the feed mainly includes diatoms, rotifers, and *Artemia*. In the postlarval stage, the feed consists of little clams, mysis, and juvenile shrimps. The usual types of diatoms include *Chaetoceros, Nitzschia, Skeletonema, Melosira*, etc. The nutritional value of diatoms, rotifers, and *Artemia* as the early diet of prawns is lower than the diet of juvenile fishes. The propagation of diatoms during culture often declines due to climatic conditions. Efforts have recently been made to use the green algae *Tetraselmis* which are always available. The nutritional value of the early diet of prawns is determined on the basis of the proteins, amino acids, sterols, lipids, and vitamins contained in the feed. Tables 9.8-9.12 show the analytical values of these components. Lab-lab is used as a natural feed in the culture regions of the Philippines. This feed is a mixture of diatoms propagated by applying inorganic fertilizer, algae, protozoans, etc. Lab-lab almost does not contain n-3 high-grade unsaturated fatty acids. However, it does contain a very small percentage of linoleic acid and linolenic acid. As a sterol, it contains 40% cholesterol.

Table 9.8: General chemical composition (%) of dry substance of feed organisms (Shirata, 1975; Watanabe, 1980)

Components	*Chaetoceros* sp.	*Tetraselmis maculata*	*Chlorella* sp.	*Brachionus plicatilis*	*Artemia* (nauplius)
Crude proteins	35.0	52.0	53.2	58.2	61.0
Crude fat	6.9	2.9	6.6	14.2	20.0
Crude ash	28.0	23.8	10.1	14.9	12.0
Carbohydrates	6.6	15.0	10.4	6.1	

Table 9.9: Amino acid composition of feed organisms (g/10 g dry sample) (Kanazawa and Teshima, 1985)

Amino acids	*Tetraselmis*	*Chlorella*, rotifers	*Artemia* (nauplius)
Aspartic acid	3.06	3.08	3.09
Threonine	1.61	1.23	1.63
Seline	1.19	1.33	1.26
Glutamic acid	3.88	3.66	4.52
Proline	2.43	1.47	4.65
Glycine	1.73	1.06	1.14
Alanine	2.39	1.30	1.45
Cysteine	0.12	0.15	1.79
Valine	2.00	1.49	1.93
Methionine	0.91	1.51	1.40
Isoleucine	1.64	1.49	1.63
Leucine	3.16	2.36	2.42
Tyrosine	1.64	1.60	1.16
Phenylalanine	2.29	1.65	1.72
Histidine	0.82	0.62	0.87
Lysine	2.44	2.61	3.72
Tryptophan	1.38	0.60	0.30
Arginine	2.19	2.00	3.23

Table 9.10: Sterol composition of feed organisms (Kanazawa *et al.*, 1971; Holden-Patterson, 1982; Teshima *et al.*, 1981; Teshima and Kanazawa, 1971)

Sterol	*Nitzschia closterium*	*Chlorella vulgaris* (IIIA)	Rotifers	Artemia	Lab-lab (Philippines)
Cholesterol			90	100	40.5
24-methylcholester-5,22-dienol	100				9.8
Ergosterol		32.80			
24-methylcholester-5,7-dienol	}				
24-methylcholester-7-enol	}	32-63			
24-E-ethyledene cholesterol			3		13.3
Cholestro-7-enol			2		
24-methylcholester-7,22-dienol			1		
24-A-ethyledene cholesterol			4		
Cholesterol					2.1
24-methylcholesterol					6.3
24-ethylcholesterol					8.5
Cholester-5,22 E-dienol					5.5
24-methylenecholesterol					2.1
24-ethylcholester-5,22-dienol					9.2

Table 9.11: Lipid composition of feed organisms (%) (Watanabe, 1980; Kanazawa, 1985; Teshima *et al.*, 1981)

Lipids	Diatoms (for *Penaeus japonicus*)	Marine *Chlorella*	Rotifers	*Artemia* (nauplius)	Lab-lab (Philippines)
14:0	23.1	4.8	—	—	3.5
16:0	9.9	20.2	16.5	11.0	36.0
16:1 n-7	14.5	29.5	16.7	4.9	14.0
18:0	0.2	tr*	3.8	5.0	2.1
18:1 n-9	2.2	8.6	15.3	30.1	3.1
18:2 n-6	1.7	4.1	6.1	8.7	6.1
18:3 n-33	0.2	tr	1.4	23.9	4.5
18:4 n-3 } 20:0 }	0.1	—	—	3.8	—
20:1	1.6	—	3.4	0.6	3.0
20:3 n-3 } 20:4 n-6 }	1.7	2.4	3.6	1.1	1.8
20:4 n-3	0.2	—	0.3	0.8	1.6
20:5 n-3	12.9	26.6	13.5	2.0	tr
22:5 n-3	—	—	6.7	—	—
22:6 n-3	—	—	2.2	—	0.3
Σn-3HUFA**	13.1	26.6	22.7	2.8	1.9

*Minute quantity.
**High-grade fatty acids.

Table 9.12: Vitamin content of feed organisms (μg/g dry substance) (Kanazawa, 1969; Morimura, 1959; Meffert and Stratmann, 1954; Hashimoto and Sato, 1954)

Vitamin	Diatoms *Chaetoceros simplex*	Green algae *Chlorella ellipsoidea*	*Scenedesmus obliquus*
Thiamine	3.15	10-23	2.7
Riboflavin	5.33	23-37	38-43
Nicotinic acid	62.30	112-125	73-107
Pantothenic acid	29.50	3.5-8.6	12-17
Vitamin B_6	1.84	0.3-2.5	1.8
Biotin	1.75	0.19-0.23	0.2
Folic acid	2.10	22-47	6.0
Vitamin B_{12}	0.0468	0.042-0.089	0.0015 (fresh material)

9.2.2 Early Microgranular Diet of Prawns

Seedling production of prawns depends on the diet organisms—diatoms, *Chlorella*, rotifers, *Artemia*, etc. For the culture of these organisms in the preparation of feed, large-scale equipment and labor are required. More-over, since the culture is greatly controlled by natural conditions, it is very difficult to ensure the required quantities of the organisms. At the same time, these organisms often undergo heavy mortality due to insufficient nutrition or a variety of pathogenic conditions. For these reasons, efforts are underway to prepare an artificial diet of high nutritional value which can effectively replace a biodiet.

Regarding the microgranular feed for prawn larvae, the following conditions are essential; the feed size should lie in the range·of 10-50 μm. Density (ability to float) may be freely adjusted. After feeding, the nutrients should not get washed away in water. After ingestion, the nutrients should get digested and be absorbed in the alimentary canal. The feed should contain a sufficient quantity of the nutritional elements required by the larvae. A microgranular feed satisfying all these conditions has now been developed.

(1) Method of preparing microgranular artificial diet

(a) *The microencapsulated diet*, MED, is a feed in which elements in the form of a solution, colloid, paste, or solid are encapsulated in a covering substance. Depending on the nature of the capsule, the feed is labeled nylon-protein MED, gelatin-gum arabic MED, or Ketosan MED. Not all MEDs contain a binder in the encapsulated feed materials and the shape and stability of the feed in water are mainly maintained by the covering coat.

The method of preparing nylon-protein MED is as follows: First, 0.5 ml span 85 is dissolved in 25 ml cyclohexan and stirred. A mixture of 2.5 ml of feed solution and 0.5 ml of diaminohexan is added and stirring continued for 3 min to ensure emulsification. The size of the microcapsule is determined by the stirring speed. If the speed is high the grains are small and if the speed is low the grains are larger. While stirring, 10 ml cyclohexan containing 0.2 ml sebacoylchloride is added in drops. Stirring is continued for another 15 min, 30 ml of cyclohexan added, and the reaction stopped. With 100 ml of cyclohexan precipitation is done twice and the precipitate (microcapsule) then washed and subjected to centrifugation at 1,000 rpm for 3 min. The separated microcapsules are collected. To wash off the solvent span 85 adhering to the microcapsules, they are scattered in 50% sucrose monolaurate-water (5-7 ml) and stirred with a magnetic stirrer for one hour. Then 100 ml water is added slowly and centrifugal separation carried out. The microcapsules precipitated are transferred in water onto a 10 μm mesh cloth for washing to remove all traces of sucrose monolaurate. The microcapsule diet is preserved in 1 mol saline and kept refrigerated.

(b) *Microbound diet*, MBD: the feed materials are bound together with a binder. The shape of the feed and its stability in water are maintained by the binder. Depending on the type of binder, the feed is classified as agar agar MBD, gelatin MBD, callagenan MBD, etc.

The callagenan MBD is prepared as follows: first, water is added to the feed material and then callagenan, while continuously stirring the mixture in a water bath at 80° C. After the callagenan has completely dissolved, the mixture is allowed to cool at room temperature. It is then dried and pulverized to a fine powder called callagenan MBD. It is preserved by refrigeration.

(c) *Microcoated diet*, MCD. Here the microbinding feed is encapsulated in a covering substance that ensures its stability in water. Depending on the type of capsule, the feed is classified as nylon-protein MCD, cholesterol lecithin MCD, etc.

In the preparation of cholesterol lecithin MCD, MBD with agar agar, gelatin, etc. is added as the binder to the cyclohexan in which cholesterol and lecithin are dissolved and continuously stirred to ensure uniform coating throughout; thereafter the cyclohexan is allowed to evaporate. The feed is used only after all traces of the solvent have been removed.

(2) Diet composition

To ascertain the protein source for the fine-grained compound diet used in the production of prawn seedlings, first the essential amio acids of the larval body protein of *Penaeus japonicus* are analyzed. A few types of protein source materials with essential amino acids close to the analyzed ones are selected. That combination of the material close to the ratio of the essential amino acids of the larval body protein is computed. Tables 9.13 and 9.14 give some examples. The composition close to the essential amino acid ratio of larval body protein is obtained by combining 37.4% brown fish meal, 17.6% yeast powder, 14.3% gluten meal, 8.8% pulverized skipjack testes, and 4.4% skim milk. To this, Alaska pollack-liver oil, soybean lecithin, cholesterol, mineral mixture, and vitamin mixture are added.

Table 9.13: Microgranular diet composition for larvae of *Penaeus japonicus*

Composition	%
Brown fish meal	37.4
Yeast powder	17.6
Gluten meal	14.3
Skipjack testes (dried and frozen)	8.8
Skim milk	4.4
Alaska pollack-liver oil	4.0
Soybean lecithin	2.0
Cholesterol	0.5
Mineral mixture	3.0
Vitamin mixture	3.0
Callagenan	5.0
Protein content	45%

Table 9.14: Amino acid composition (g) of diet protein and *Penaeus* larvae

Amino acids	Amino acid contents in 45 g of bodyprotein of *Penaeus* larvae (P_1)	Amino acid contents in 45 g of protein of microgranular diet (brown fish meal + yeast powder + gluten meal + skipjack testes powder)
Arginine	3.66	3.68
Histidine	1.15	1.11
Isoleucine	2.28	2.24
Leucine	3.02	3.08
Lysine	3.31	3.28
Methionine	1.46	1.44
Phenylalanine	2.23	2.24
Threonine	1.45	1.47
Tryptophan	1.67	1.68
Valine	2.14	2.19

(3) Method of diet supply

The microgranular diet is supplied in differing amounts to *Penaeus* larvae at various stages of growth: less than 53 μm for zoea 1; 53-125 μm for zoea 2—mysis 1; 125-250 μm for mysis 2—postlarva 2; and 250-350 μm for larvae beyond postlarvae 3. The quantity provided is as follows: 0.16 mg/larva/day for zoea 1-3; 0.20 mg/larva/day for mysis 1-3; 0.24 mg/larva/day for postlarva 1-3; and 0.30 mg/larva/day for postlarvae 4 and beyond. The number of feeds is six per day. The water change is 10-40%/day. The microgranular diet is not scattered in one lot but in small installments over a wide area on the water surface. It is effective to use an automatic supplying machine. The pH and ammonia content of the rearing water should be monitored daily and a suitable change of water carried out.

(4) Results of rearing

a) Practical example 1: Small-scale seedling production was carried out in which 8,000 larvae were grown in a 0.5-ton water tank from zoea 1 up to postlarva 6. The results are given in Table 9.15. Even with a microgranular artificial diet alone the larvae grew steadily and at postlarva 6 the survival was as high as 90%.

b) Practical example 2: 300,000 just-hatched larvae of *Penaeus japonicus* were released in a 16-ton water tank. The larvae were split into two groups with one group given biofeed (*Chaetoceros, Artemia,* and commercial combined diet) and the other only callagenan microgranular feed. The two groups were compared (Fig. 9.7). The results indicate that even when fed on a microgranular artificial diet only, the larvae grew steadily through zoea 1, 2, and 3; mysis 1, 2, and 3; and attained the postlarval stage. The survival at postlarva 8 was 70%.

Table 9.15: Small-scale seedling production of *Penaeus* larvae using a microgranular diet (0.5-ton water tank)

Duration of rearing (days)	No. of larvae	Survival (%)	Developmental stages*
1	80,000	100	$Z_1 = 100$
2	80,000	100	$Z_2 : Z_3 = 90 : 10$
3	80,000	100	$Z_3 = 100$
4	—	—	—
5	80,000	100	$M_1 : M_2 = 80 : 20$
6	—	—	—
7	80,000	100	$M_2 : M_3 = 30 : 70$
8	80,000	100	$P_1 = 100$
9	80,000	100	$P_2 = 100$
10	80,000	100	$P_3 = 100$
11	80,000	100	$P_4 = 100$
12	72,000	90	$P_5 = 100$
13	72,000	90	$P_6 = 100$

*Figures indicate percentage.

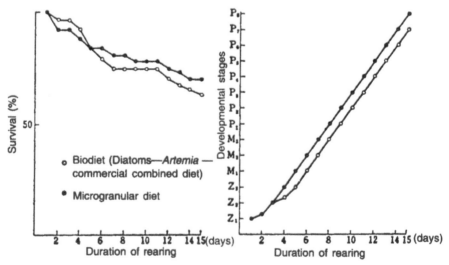

Fig. 9.7. Quantitative production of seedlings of *Penaeus japonicus* using a microgranular diet (16-ton water tank).

The particulate sizes of the microgranular diet range from 10-350 μm. If the pellet coat or binder is insufficient, upon suspension the nutritional components dissolve in the water and pollution often results. On the other hand, if these are too strong the larvae experience difficulty in digesting the microparticulates after ingestion. The enzymatic activity for breaking down the nutrients is particularly weak in the digestive tract of zoea and mysis. Hence the components must be soluble low molecular compounds that are readily digestible.

Details of the feeding behavior of larvae have yet to be clarified. To improve feeding efficiency the capsule of the microgranular feed requires modification. Concomitantly the structure of the rearing tank should be altered to better accommodate artificial feeding. To ensure water quality, cleansing of the rearing tanks should be automated. Sooner or later, microgranular artificial feed will replace biofeed in the early period of rearing *Penaeus japonicus* and will probably find wide application in the culturing of other shrimp species throughout the world.

9.2.3 Combined Diet for Prawn Culture

As a natural feed for prawn culture, small clams, juvenile shrimps, juvenile crabs, mysids, krills, and juvenile fishes were once used. However, considering the high cost of clams, ease of handling biofeed, and concomitant prevention of water pollution, a pellet form of a combined diet was developed and became popular. Initially, growth on this combined diet was not very good. Once the nutritional requirements of *Penaeus japonicus* became clear, however, it was possible to develop a suitable combined diet which results in steady growth and high survival rate. Among the various prawn culture regions of the world, there are places where culture depends upon natural feed. Starting with *Penaeus* culture in Japan, other regions in the world started shifting to the pellet form of combined diet for prawn culture. Production has proven to be very high in relation to unit area of culture.

(1) Materials used in diet

a) Protein source: An animal source is usually high in protein content and moreover the balance of amino acids constituting the protein is also very good. Therefore animal protein is used as the protein source material. The protein content and amino acid balance of white fish meal in particular is very good. Fish meal is safe to use. Brown fish meal, processed with various types of sardines and mackerel, compared to white fish meal, is inferior in feed efficiency. Moreover, it poses the problem of lipid deterioration. Recently, the preparation technique was greatly advanced and it is now possible to prepare good quality brown fish meal.

Skim milk, shrimp meal, and krill meal are also found to be rich in good-quality protein.

Squid meal is an excellent protein source for prawns and is widely used in the preparation of a combined diet. However, since the catch rate of squids has recently declined, and the price has also increased, prawn feed manufacturers are trying to find an alternative protein source.

Wheat gluten, corn gluten, and fat-free soybean, used as a plant protein source, lack one or two essential amino acids. However, they are widely used as relatively good proteins. So too are microorganisms such as yeast and *Spirulina*. In any case, the selection of protein source should be made after analyzing the amino acids of prawns and then selecting the type of protein which is close to the amino acid composition of the prawns, both in terms of quantity and ratio. Such proteins will have an encouraging effect both for growth and survival.

The amino acid composition of the important protein sources is given in Table 9.16.

Table 9.16: Amino acid composition of feed components (g/100 g dry sample)

Amino acids	Squid meal	Krill meal	Brown fish meal	White fish meal	Skipjack ovaries	Skipjack testes	Fazer meal	Soy- bean meal	Gluten meal	Brauer yeast
Essential										
Methionine	2.80	0.69	1.33	1.88	1.85	0.96	0.51	1.14	0.86	0.83
Threonine	2.22	1.61	2.35	2.30	2.46	1.84	3.06	2.72	1.48	2.09
Valine	2.45	2.25	2.63	3.00	3.24	2.33	5.44	3.10	2.12	2.45
Isoleucine	2.60	2.35	2.21	2.66	2.76	1.77	3.68	3.29	1.99	2.09
Leucine	4.07	3.03	4.04	4.50	4.54	3.07	5.81	5.49	7.53	2.98
Phenylalanine	2.34	2.21	2.44	2.59	2.70	1.79	3.75	4.61	3.11	2.05
Histidine	1.14	1.00	1.92	1.34	1.74	0.97	0.66	1.86	0.96	1.01
Lysine	4.47	3.83	5.16	5.66	4.68	3.26	1.18	5.30	0.84	3.28
Tryptophan	1.07	0.51	2.52	1.63	1.98	0.87	1.15	3.03	0.38	0.25
Arginine	4.56	2.96	3.39	4.26	4.44	7.76	5.65	5.75	1.70	3.10
Nonessential										
Aspartic acid	5.66	4.51	5.05	5.66	4.49	3.49	4.70	8.14	2.83	4.38
Serine	1.82	1.23	1.70	1.87	2.19	1.54	5.62	2.71	1.76	2.20
Glutamic acid	7.99	5.35	6.47	8.83	6.98	4.92	8.09	12.71	10.47	6.57
Proline	4.02	2.65	2.78	2.66	3.33	2.55	7.71	4.26	3.99	1.89
Glycine	2.12	1.59	2.85	2.56	1.93	2.19	4.19	2.14	1.03	2.00
Alanine	2.54	1.88	3.08	2.96	3.19	2.11	2.67	2.62	3.66	2.76
Cysteine	0.43	0.73	0.37	0.40	0.93	0.36	2.61	0.40	1.07	0.67
Tyrosine	2.10	1.82	2.13	2.34	2.68	2.06	2.43	3.12	2.67	1.91
Taurine	2.34	0.48	1.11	0.56	1.57	3.35	—	—	—	—
Total	56.74	40.68	53.53	57.66	57.59	47.19	69.54	72.39	48.45	42.51

b) Lipid source: The essential lipids of *Penaeus japonicus* have been worked out and therefore it is only necessary to add the oil or fat which contains these lipids. Table 9.17 gives the lipid composition of Alaska pollack-liver oil in a refined form (feed oil). Feed oil contains eicosapentanoic acid and decosahexanoic acid which are n-3 high-grade unsaturated fatty acids and also essential for *P. japonicus*. Soybean oil mainly contains linoleic acid. The diet efficiency and growth of *P. japonicus* are found to be very high when Alaska pollack-liver oil and soybean oil are added in mixed form in the ratio of 3:1 or 1:1. As for other types of prawns, there may be some variation in requirement for essential fatty acids.

Table 9.17: Lipid composition of refined liver oil of Alaska pollack (Uehara, 1978)

Lipids	%	Lipids	%
12:0	0.04	20:-2n-6	0.27
14:0	6.19	20:3 n-3 ⎫	
15:0	0.27	20:4 n-6 ⎬	0.22
16:0	10.13	20:5 n-3	9.86
16:1 n-9 or 7	8.31	20:4 n-3	0.42
18:0	1.92	22:1	14.83
18:1 n-9 or 7	14.01	22:5 n-3	0.71
18:2 n-6	0.67	22: 5 n-6	0.18
18:3 n-3	0.26	22:6 n-3	4.23
18:3 n-6	0.15	unknown	4.18
18:4 n-3 ⎫			
20:0 ⎬	1.61	Σ n-6	1.3-1.5
20:1 n-9 or 7	20.54	Σ n-3	17.0-17.2

c) Other additions to the diet: In the combined diet of prawns, it is necessary to fortify the nutrients with vitamins and minerals. An oxidation-resistant agent and coloring agent are also added in some cases.

Like fishes, prawns do not feed soon after the feed is supplied. Moreover, they approach the feed cautiously and ingest it by nibbling. Therefore it is necessary to add a binder to the diet. The feed should be provided with a suitable cohesive to facilitate prawn feeding and also to prevent dissolution of the nutritional elements in the water. Water pollution is thereby likewise prevented. Active gluten, wheat flour, and starch are used as binding agents.

(2) Composition of combined diet

The major item, both in terms of quantity and economy, in the combined diet of prawns is protein. To determine the protein source, the ratio of essential amino acids is worked out first on the basis of amino acid analysis of the prawn muscles. In many cases, the required quantities of essential amino acids are not clear. Therefore, these are decided with the value of the concentration of individual essential amino acids in the prawn protein taken as the standard.

In one of the combined diet compositions for *Penaeus japonicus*, the protein source is squid meal and soybean as both are rich in protein. The arginine source is skipjack testes, the lysine source fish meal and yeast, and the methionine source krill meal. The ratio of combination is calculated by considering both the ratio (%) and quantity of essential amino acids. Computational results are given in Tables 9.18-9.20.

The composition of the combined diet of *Penaeus vannamei* used in experiments at Texas A and M University (Lawrence *et al.*, 1985) is given in Table 9.21.

Table 9.18: Amino acid composition of *Penaeus japonicus* and diet (g/100 g dry sample)

Amino acids	*P. japonicus*	Diet 1	Diet 2
Methionine	1.22	1.33	1.26
Threonine	1.42	1.83	1.76
Valine	1.76	2.07	2.09
Isoleucine	1.76	1.88	1.86
Leucine	2.99	3.23	3.18
Phenylalanine	1.56	1.91	1.93
Histidine	0.84	0.94	1.10
Lysine	3.19	4.01	3.65
Tryptophan	1.54	1.28	1.44
Arginine	4.30	4.49	4.35
Aspartic acid	3.89	3.89	4.05
Serine	1.03	1.40	1.43
Glutamic acid	6.35	5.60	5.74
Proline	3.20	2.08	2.35
Glycine	3.17	1.78	1.66
Alanine	2.09	2.07	1.99
Cysteine	0.21	0.35	0.33
Tyrosine	1.48	1.86	1.83
Total	42.00	42.00	42.00

Table 9.19: Ratio of individual essential amino acids with methionine as 1.00

Amino acids	*Penaeus japonicus*	Diet 1	Diet 2
Methionine	1.00	1.00	1.00
Threonine	1.16	1.38	1.40
Valine	1.44	1.56	1.66
Isoleucine	1.44	1.41	1.48
Leucine	2.46	2.43	2.52
Phenylalanine	1.28	1.44	1.53
Histidine	0.69	0.71	0.87
Lysine	2.62	3.02	2.90
Tryptophan	1.27	0.96	1.14
Arginine	3.53	3.38	3.45

Table 9.20: Composition of combined diet of *Penaeus japonicus* (%)

Composition	Diet 1	Diet 2
Soybean meal	—	14.9
Squid meal	19.8	—
Krill meal	—	13.5
White fish meal	20.3	10.2
Brown fish meal	—	16.0
Yeast powder	15.1	—
Skipjack testes	25.4	25.4
Alaska pollack-liver oil	4.0	4.0
Soybean lecithin	3.0	3.0
Cholesterol	0.5	0.5
Mineral mixture	5.0	5.0
Vitamin mixture	6.0	6.0
Cellulose	0.9	1.5

Table 9.21: Composition of combined diet of *Penaeus vannamei* (Lawrence *et al.*, 1985)

Diet components	Diet composition (%)					
	A			B		
	36A	29A	22A	36B	29B	22B
Shrimp meat	36.00	30.90	20.60	29.40	21.40	13.40
Menhaden meal	3.50	3.50	3.50	3.20	3.20	3.20
Squid meal	2.00	2.00	2.00	1.50	1.50	1.50
Fish soluble	2.00	2.00	2.00	2.00	2.00	2.00
Rice bran	22.75	35.00	36.00	35.00	35.50	34.50
Corn starch	12.25	14.40	25.00	7.20	17.50	29.20
Soybean meal	3.00	3.00	0.25	6.50	4.00	2.20
Wheat gluten	6.00	0.00	0.00	6.25	4.25	2.00
Casein (vitamin-free)	3.30	0.00	0.00	0.00	0.00	0.00
Cellulose	1.00	0.90	1.85	1.00	1.75	2.70
Vitamin mixture	2.00	2.00	2.00	2.00	2.00	2.00
Mineral mixture	1.00	1.00	1.00	1.00	1.00	1.00
Pollack-liver oil	0.70	3.00	0.80	0.45	0.90	1.30
Lecithin	1.00	1.00	1.00	1.00	1.00	1.00
Cholesterol	0.50	0.50	0.50	0.50	0.50	0.50
Sodium hexametaphosphite	1.00	1.00	1.00	1.00	1.00	1.00
Argenic acid	2.00	2.50	2.50	2.00	2.50	2.50

A—animal protein: plant protein (2 : 1)
B—animal protein: plant protein (1 : 1)

(3) Results of rearing

a) Practical example (Penaeus japonicus): The results of rearing on a combined diet carried out at the Fisheries Experimental Station of Kagoshima prefecture (Deshimaru, 1981) are given in Tables 9.22 and 9.23.

b) Practical example (Penaeus monodon and others): A comparison of the growth of cultured prawns such as *Penaeus monodon, P. stylirostris, P. pencillatus, P. japonicus, P. semisulcatus* and *Metapenaeus* sp. studied by Liao (1981) at the Fisheries Experimental Center of Taiwan (Fig. 9.8) showed the growth of *P. monodon* to be the most rapid.

c) Practical example (Penaeus stylirostris): Table 9.24 shows the composition of the combined diet on which *P. stylirostris* was reared at Texas A and M University (Lawrence *et al.*, 1981). The results of testing the control effect of replacing shrimp meal with yeast indicated that with the combination of shrimp meal 15% + yeast 15%, weight gain increased to 137% and survival to 89%.

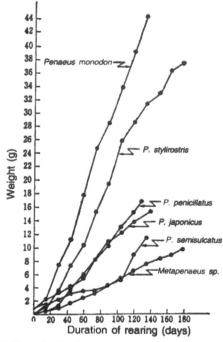

Fig. 9.8. Growth of cultured prawns in Taiwan (Liao, 1981).

Table 9.22: Composition of diet of *Penaeus japonicus* (%) (Yoshimaru, 1981)

Diet components	Diet			
	102	10~	104	105
Main components				
Squid meal	4.0	1.0	49.0	2.0
Shrimp meal	3.0	2.0	2.0	10.0
Testes meal[1]	23.0	10.0	6.0	2.0
Yeast	22.0	45.0	1.0	2.0
Casein	18.0	16.5	9.0	45.0
Gluten	5.0	5.0	5.0	5.0
Subtotal	75.0	79.5	72.0	66.0
Additives				
Phospholipids[2]	2.0	—	1.0	—
Mixed oil[3]	—	3.5	0.6	5.3
Cholesterol	—	0.8	—	1.0
Glycogen[4]	8.0	6.5	9.0	9.0
$NaH_2PO_4 \cdot 2H_2O$	5.0	4.0	7.5	8.0
KCl	—	—	1.0	2.0
Vitamin mixture	4.0	4.0	4.0	4.0
Cellulose	6.0	1.7	4.9	4.7
Subtotal	25.0	20.5	28.0	34.0
Total	100.0	100.0	100.0	100.0

[1] Skipjack.   [2] Squid   [3] Alaska pollack-liver oil : soybean oil (2 : 1).   [4] Oyster.

Table 9.23: Results of rearing *Penaeus japonicus* on a combined diet[1] (Deshimaru, 1981)

Items	Little clam	Diet					
		102	103	104	105	A[2]	B[2]
No. of prawns							
Before rearing	25	25	25	25	25	25	25
After rearing	24	22	20	22	20	24	25
Average weight (g)							
Before rearing	0.59	0.58	0.58	0.58	0.60	0.58	0.59
± S.D.	±0.60	±0.05	±0.05	±0.05	±0.06	±0.05	±0.06
After rearing	2.35	2.29	1.85	1.60	1.87	1.38	1.31
± S.D.	±0.50	±0.37	±0.27	±0.29	±0.30.	±0.37	±0.37
Weight gain (%)	298.3	294.8	219.0	175.9	211.7	137.9	122.0
Diet efficiency (%)	98.2	94.6	47.1	95.3	67.5	71.6	83.7

[1] Rearing temperature, 20-23° C, rearing period, 30 days.
[2] Commercial combined diet.

Table 9.24: Composition of combined diet and results of rearing *Penaeus stylirostris* (Lawrence *et al.*, 1981)

Diet components	Diet composition (%)				
	25	26	27	28	30
Shrimp meal	26.5	15.7	5.0	—	31.5
Yeast	5.0	15.8	26.5	31.5	—
Soybean meal	3.0	3.0	3.0	3.0	3.0
Fish meal	8.0	8.0	8.0	8.0	8.0
Rice bran	44.0	44.0	44.0	44.0	29.0
Squid meal	5.0	5.0	5.0	5.0	5.0
Corn meal	—	—	—	—	15.0
Fish soluble	2.0	2.0	2.0	2.0	2.0
Vitamin mixture	2.0	2.0	2.0	2.0	2.0
Arginine	2.5	2.5	2.5	2.5	2.5
Sodium hexametaphosphite, Na	1.0	1.0	1.0	1.0	1.0
Lecithin	1.0	1.0	1.0	1.0	1.0
Protein (%)	34.3	35.8	33.9	31.5	32.9
Lipid (%)	4.2	5.6	6.4	5.8	5.7
Carbohydrate (%)	20.6	25.1	28.7	25.0	35.0
Water (%)	6.1	5.8	4.1	3.1	4.4
Weight gain (%)	126	137	93	88	116
Survival (%)	75	89	82	91	82

REFERENCES

Association of Fisheries of Japan (Nihon Suisangakkai). 1978. Fisheries Series, no. 22. Fish Culture and Diet Fat. Koseisha-koseikaku.

Boghen AD, Castell JD. 1981. Nutritional value of different dietary proteins to juvenile lobsters (*Homarus americanus*). Aquaculture, 22.

Deshimaru K. 1981. Study of nutrition and diet of *P. japonicus*. Manual of Fisheries Experiment Station, Kagoshima Prefecture, 12.

Fenucci JL, Lawrence AL, Zein-Eldin ZP. 1981. The effect of fatty acid and shrimp meal composition of prepared diets on growth of juvenile shrimp (*Penaeus stylirostris*). J. World Maricul. Soc., 12.

Kanazawa A. 1980. Nutrient requirements of prawns. Ocean Signs (Kaiyokagaku), 12 (12).

Kanazawa A. 1981. In: Proc. 2nd Int. Conf. on Aquaculture Nutrition. GD Pruder, C Langdon, DE Conklin (eds.). Louisiana State Univ.

Kanazawa A. 1983. The effect of phospholipid on marine organisms. Report of Fat Research Association, B. series, no. 18.

Kanazawa A. 1985. Microgranular artificial diet as early diet for *P. japonicus*. Aquaculture (Yoshoku), 22 (2).

Kanazawa A, Teshima S. 1981. Essential amino acids of the prawn. Bull. Japan, Soc. Sci. Fish., 47.

Kanazawa A, Teshima S. 1983. Development of microgranular artificial diet for juvenile fishes, crustacean larvae and molluscan larvae. Aquaculture (Yoshoku), 20 (11).

Kanazawa A, Teshima S, Matsumoto S, Nomura T. 1981. Dietary protein requirement of the shrimp *Metapenaeus monoceros*. Bull. Japan, Soc. Sci. Fish., 47.

Kanazawa A, Teshima S, Sasada H, Abdel-Rahman S. 1982. Culture of the prawn larvae with microparticulate diets. Bull. Japan, Soc. Sci. Fish., 48.

Kean JC, Castell JD, Trider DJ. 1985. Juvenile lobster (*Homarus americanus*) do not require dietary ascorbic acid. Can. J. Fish. Aquat. Sci., 42.

Kean JC, Castell JD, Boghen AD, d'Abramo LR, Conklin DE. 1985. A re-evaluation of the lecithin and cholesterol requirements of juvenile lobster (*Homarus americanus*) using crab protein-based diets. Aquaculture, 47.

Kome K. 1985. Fisheries Series, no. 54. Fish Culture Diet. Koseisha-koseikaku.

Shigeno S (ed.) 1980. Nutrition and Diet of Fishes. Koseisha-koseikaku.

Shigueno K. 1975. Shrimp Culture in Japan. Assoc. Int. Tech. Promotion.

Shirata S. 1975. Fisheries Diet Biology. Koseisha-koseikaku.

Smith LL, Lee PG, Lawrence AL, Strawn K. 1985. Growth and digestibility by three sizes of *Penaeus vannamei* Boone: Effects of dietary protein in level and protein source. Aquaculture, 46.

Teshima S. 1984. Nutritional requirement of *Penaeus* larvae. Report of Fat Research Association, B. series, no. 19.

Teshima S, Kanazawa A. 1983. Effects of several factors on growth and survival of prawn larvae reared with microparticulate diets. Bull. Japan, Soc. Sci. Fish., 49.

Teshima S, Kanazawa A. 1984. Effects of protein, lipid, and carbohydrate levels in purified diets on growth and survival rates of prawn larvae. Bull. Japan, Soc. Sci. Fish., 50.

Teshima S, Kanazawa A, Sakamoto M. 1982. Microparticulate diets for larvae of aquatic animals. Min. Rev. Data File. Res. 2.

Teshima S, Kanazawa A, Sasada H. 1983. Nutritional value of dietary cholesterol and other sterols to larval prawn, *Penaeus japonicus* Bate. Aquaculture, 31.

Teshima S, Kanazawa A, Sasada H, Kawasaki M. 1982. Requirements of the larval prawn, *Penaeus japonicus*, for cholesterol and soybean phospholipids. Mem. Fac. Fish., Kagoshima Univ., 31.

Species Index